The Institution of
StructuralEngineers

The regenerative structural engineer

This version (1.0) published January 2024 by
The Institution of Structural Engineers
International HQ
47–58 Bastwick Street
London, EC1V 3PS
United Kingdom
www.istructe.org

Typeset by Riverside Publishing Solutions Ltd

Printed and bound by Hobbs The Printers Ltd

WORLD LAND TRUST™
www.carbonbalancedprint.com
CBP2250

MIX
Paper | Supporting responsible forestry
FSC® C020438
FSC www.fsc.org

ISBN (print) 978-1-906335-68-7
ISBN (pdf) 978-1-906335-69-4

Contents

Acknowledgements

The Institution would like to thank Jenny Spratt (J.Spratt Design) for creating the visual approach for this book.

Permission to reproduce the following has been obtained, courtesy of these individuals/organisations:

Full page imagery:
Page xii: ©James McKay from the book *Unlocking Sustainable Cities* by Paul Chatterton, Pluto Press
Page 6: Lucy Rees
Page 70: MASS Design Group

Figures
Page x: ©Zach Norman
Figure 2.1 (right): ©McGee
Figure 2.6: Wikipedia Commons ©DoughnutEconomics (CC BY-SA 4.0)
Figure 3.1: Derived/adapted from H3Uni (CC BY-SA 4.0)
Figure 3.4: Wikipedia Commons ©Schmiebel (CC BY-SA 4.0)
Figure 3.5: ©The Trustees of the Natural History Museum, London
Figure 3.6: Derived/adapted from *The Nature and Properties of Soils* (15th edition) by R Weil and N Brady (Pearson, 2017)
Figure 3.7: Wikipedia Commons ©Ingmar Runge (CC BY-SA 3.0)
Figure 3.11: ©Ben Law
Figure 3.12: ©Jamie Miller (Tŷ Pren) and Maya Burchell Haslett (Tŷ Pren)
Figure 4.2: ©George Fereday
Figure 4.3: ©On Screen Creations
Figures 4.5, 4.15 and 6.9: ©Buro Happold
Figure 4.6: ©Raffaele De Risi

Figure 4.7: ©Format Engineers Ltd
Figure 4.8: ©EA Piling
Figure 4.9: ©Heritage Foundation of Pakistan
Figures 4.11, 4.12 and 9.7 (left): ©Heyne Tillett Steel
Figure 4.13: ©Margaret Cooke, Integral Engineering Design Ltd (Reproduced by kind permission of the Chapter of Wells Cathedral)
Figure 4.14: ©Nederlandse Bruggenstichting
Figure 4.16: ©WeCanMake/Ibolya Feher (Photographer)
Figure 4.17: Derived/adapted from Csikszentmihalyi M. *Implications of a systems perspective for the study of creativity* in: Sternberg, R.J. (Ed.), *Handbook of Creativity* (Cambridge University Press, 1999)
Figure 4.18: ©IBUKU/Tomasso Riva [The Arc] and Jan Friedlein/AKT II [HYLO]
Figure 4.19: ©Alexie Sommer/Constructivist
Figures 4.20, 4.21 and 6.5: ©YesmakeLDN
Figures 4.22, 4.23 and 5.4–5.7: ©MASS Design Group
Figure 5.3: ©Iwan Baan
Figures 5.9–5.12: ©Agile Homes
Figure 5.13: ©Paul Jaquin/eZED
Figures 5.14–5.18: ©BASE
Figure 5.19: ©David Grandorge
Figures 5.20–5.22 and 9.5 (middle): ©Ellen Grist
Figures 5.22 (right): ©Henrietta Williams
Figure 6.8: ©Katie Symons
Figure 6.10: ©Smith and Wallwork
Figure 6.13: ©Will Arnold
Figure 6.14: Derived/adapted from *The Cultural Web* by MindTools
Figure 7.2: ©Tom Clark
Figure 7.4: ©Alexie Sommer
Figure 9.3: ©Fast + Epp
Figure 9.4: ©James McKay from the book *Unlocking Sustainable Cities* by Paul Chatterton, Pluto Press
Figure 9.5 (bottom): ©James McKay *Dreams of a Low Carbon Future*
Figure 9.6: ©Integral Engineering Design Ltd

The authors

OLIVER BROADBENT

1851 Fellow in Regenerative Design

Oliver in the process of writing this book

Oliver has a diverse background — he studied Chemistry at The University of Oxford, Civil Engineering at Imperial College London, and worked as an engineer at Expedition. However, he quickly diversified again, helping to establish Think Up, the design education arm of the Useful Simple Trust[0.1]. Alongside other initiatives, Think Up produced the highly influential *Embedding Sustainability in the Undergraduate Civil Engineering Curriculum*[0.2] back in 2012 — when many organisations were just waking up to sustainability. A quick flick through the guidance now and you will spot, if you have any familiarity with the concepts of regenerative design, a number of regenerative ideas.

In 2018 Oliver left Think Up to start his own company Constructivist[0.3]. From here he has continued the conversation on sustainability, regenerative design, the climate and biodiversity emergency, and creativity. He has run training on what to do after declaring a climate emergency and continues to run training on both net-zero design and conceptual design with the Institution of Structural Engineers. In 2020, Oliver was honoured to receive the Sir Misha Black Award for Innovation in Design Education. In 2022 he established the Constructivist Regenerative Design Lab and later that year was appointed 1851 Fellow in Regenerative Design.

JAMES NORMAN
Professor of Sustainable Design

James in the process of writing this book

James has been a structural engineer all his life. He studied Civil Engineering at the University of Nottingham, joined Whitbybird in 2000, left to do a PhD at the University of Bristol in structural dynamics between 2003–6, and rejoined Whitbybird (who became Ramboll soon after) in 2006. James however, continued to teach structural design one day per week. In 2011 he started teaching a unit on sustainable construction and in 2012 joined Integral Engineering Design to get more hands-on experience in sustainable design (including rammed earth, straw-bale, timber and reuse).

James has designed using a number of different materials including steel, concrete, timber, masonry, earth, straw and glass. He has worked on a wide variety of projects; from the façade of the Tate Modern extension to a small glass table at the Tate St Ives.

In 2015 James became a full time academic but continues to think and behave like a structural engineer. He teaches structural design to years 2–4 for half the year and during the other half he writes books and articles for structural engineers including *Conceptual design of buildings* (IStructE)[0.4], *Structural Timber Elements: A pre-scheme design guide* (BM TRADA)[0.5] and *Designing timber structures: An introduction* (BM TRADA)[0.6]. In 2020 James was awarded a National Teaching Fellowship and became Professor of Sustainable Design.

REFERENCES

0.1 Useful Simple Trust. *Purpose-driven design for our changing environment.* Available at: www.usefulsimple.co.uk [Accessed: December 2023].

0.2 Think Up. *Embedding sustainability in the undergraduate civil engineering curriculum.* Available at: https://thinkup.org/reports/embedding-sustainability-in-the-undergraduate-civil-engineering-curriculum/ [Accessed: December 2023].

0.3 Constructivist. Available at: https://constructivist.co.uk/ [Accessed: December 2023].

0.4 The Institution of Structural Engineers. *Conceptual design of buildings.* London: IStructE Ltd, 2020.

0.5 Norman J. *Structural Timber Elements: A pre-scheme design guide.* (2nd edition). High Wycombe: BM TRADA, 2017.

0.6 Norman J and Thompson A. *Designing timber structures: An introduction.* High Wycombe: BM TRADA, 2020.

CHAPTER 1

Introduction

What if, in our work designing buildings, towns and cities, we could not only minimise harm to the environment but actively contribute to its healing and regeneration?

The concept of regenerative design started to gain prominence with built environment professionals in the UK around the time of the IPCC's *Special Report: Global Warming of 1.5°C*[1.1] in 2018. That report triggered a growing realisation among people in the built environment sector, including many structural engineers, that despite striving for sustainability for more than 20 years, our industry was not doing enough to limit its impacts on the climate and on the ecosystem that supports life on Earth. Within the global structural engineering community, this realisation manifested itself in initiatives like 'UK Structural Engineers Declare a Climate & Biodiversity Emergency'[1.2], which had been signed by 270 structural engineering firms at the time of writing.

As our industry begins to confront the stark reality of climate breakdown and biodiversity loss, it becomes clear that a profound transformation is needed in the built environment sector. This transformation requires us to rethink not only our approach to individual projects but also how we organise our industry and the mindsets through which we see our work. The Structural Engineers Declare statement acknowledges that 'together with our clients, we will need to commission, design and build cities and infrastructure as indivisible components of larger, constantly regenerating and self-sustaining systems in balance with the natural world'.

Embracing this approach to design will require us to move away from a model of continuous growth towards a more holistic one. Design recognising that everything we produce is part of a wider living system that must remain in balance. Design that not only repairs the damage done to our climate and ecosystems but that enables all life, humans included, to flourish. In other words, an approach to design that is **regenerative**.

The goal of regenerative design is for human and living systems to survive, thrive and co-evolve[1.3].

But what does a regenerative approach to design look like in reality? How is it different from the sustainability approaches that we have been taking so far and what role should structural engineers play? As the term 'regenerative' gains traction, it is becoming increasingly important to ask these questions, both so we can make the most of what this regenerative framing offers, and so that we can avoid the term becoming subverted into marketing jargon.

ABOUT THIS BOOK

Split into four parts, this book explores what it means to be a regenerative structural engineer. By asking this question we aim to provide purpose and motivation for structural engineers at all stages in their careers as they think about the role they can play in responding to the climate and biodiversity emergency.

At the same time, we acknowledge that not all engineers may feel the relevance of regenerative design in their daily work. But while it may not be expressed in such terms, we predict that as the climate and biodiversity emergency deepens, more and more clients will be asking design teams to respond to complex questions about our wider ecological impact as well as our carbon impact. Our intention is that this exploration of regenerative structural engineering can provide a framing that helps answer these questions.

Here is a summary of the content that follows:

Part 1 (Chapters 2–3): Introducing regenerative thinking

- Exploring regenerative design through a systems lens
- How regenerative design goes beyond sustainable design
- Plotting a path from our current way of working to a regenerative industry
- Taking inspiration from thriving living systems for how we should be designing

Part 2 (Chapters 4–6): Creating the transition

- How can we design our projects more regeneratively?
- How can we enable the industry to operate more regeneratively?
- How can we enable engineers to think more regeneratively?

Part 3 (Chapters 7–9): Becoming regenerative

- How can we create and sustain regenerative practice?
- The role of the regenerative structural engineer
- Practicing place-based continuous design (and an artefact from the future)

Part 4: Appendices

- The 12 Levers of Systems Change
- Definitions of 'regenerative design'
- Creating your own artefact from the future
- Bibliography

REFERENCES

1.1 Intergovernmental Panel on Climate Change. *IPCC Special Report: Global Warming of 1.5°C.* Available at: www.ipcc.ch/sr15 [Accessed: December 2023].

1.2 Engineersdeclare.com. *UK Structural Engineers Declare Climate & Biodiversity Emergency.* Available at: www.structuralengineersdeclare.com/ [Accessed: December 2023].

1.3 Constructivist. *The goal of regenerative design.* Available at: https://constructivist.co.uk/the-goal-of-regenerative-design/ [Accessed: December 2023].

Part 1
Introducing regenerative thinking

In Chapter 2 we look at the need for a shift from 'sustainable' to 'regenerative' thinking. We characterise regenerative design using a simplified systems model.

In Chapter 3 we begin to explore what it would take to shift our construction industry towards more regenerative ways of working.

These chapters lay the groundwork for Part 2 — where we explore how we can practice more regeneratively in our work as structural engineers.

CHAPTER 2

From sustainable to regenerative

Why do we need to think regeneratively? To help us answer this question we begin by providing a simple systems framework; an approach that might be familiar to some but might provide a new angle for others. Through this systems perspective, we explore why our industry's ambition to be sustainable has, despite best intentions, failed to protect the physical and living systems — air, water, soil, climate and ecology — that we depend on to survive.

We then go on to introduce regenerative design; an approach to design in the built environment that is emerging in response to the climate and biodiversity emergency. Using our systems framework, we will see how regenerative design seeks to go much further than sustainability — envisioning a future where integrated human and living systems are not only repaired, but have recovered their capacity for self-repair.

In writing this chapter we have assumed no previous knowledge of systems thinking; however, if you would like to delve deeper into this subject then we strongly recommend reading Donella Meadows' short primer *Thinking in Systems*[2.1]. We provide built environment examples of her '12 levers' (which informs much of our content) in Appendix A.

2.1 A SYSTEMS VIEW OF CONSTRUCTION

> "Systems thinking is a framework for seeing interrelationships rather than things; for seeing patterns of change rather than static snapshots."[2.2]

Systems thinking is a rigorous approach that is used in many fields; from medicine to planning NASA missions. While not a typical tool for most structural engineers, systems thinking enables us to see the impact of our design choices on the wider construction industry, and in turn on the wider ecosystem.

For some, seeing the construction industry as a system will be intuitive. For others, a bit more exploration would be helpful.

2.1.1 A construction site as system

We can think of any collection of interrelated components as operating as a system. Medicine is based on the idea that every part of the body is interconnected. Rather than seeing the body as a series of separate components, medicine sees the body as a cohesive whole, and tries to understand the interrelationships between these components over time. This systems-orientated perspective aims to achieve better health outcomes by recognising that all components of the system are interconnected.

Similarly, we can think of a construction site as a system.

The first step in describing any system is to define its boundaries. In a simple view of the human body, the boundary of the system is the skin. In our systems model of a construction site, we'll set the system boundary as the red line that describes the site boundary on drawings. Everything that exists within that site boundary is part of the construction site system (Figure 2.1).

Figure 2.1: Two possible system boundaries: the skin of a human and the boundary of a construction site

In the human body, the input is nutrients, water and air. Processing all those nutrients are a network of organs that turn these inputs into the structure of the body. On a construction site, the inputs are the construction materials. The architects, engineers and contractors, together with their machinery are like the organs that convert these inputs into the structure of the building.

Linking the organs to the rest of the body, communicating information, controlling action and transmitting nutrients are the arteries, veins and nervous system. On a construction site, risk assessments and method statements guide the actions of individuals and teams. Information is transmitted through various means, such as specification documents, building models or drawings, and visual inspections.

Just as changing medication for one part of the body will affect the rest, changing an aspect of the construction process is likely to have knock-on effects on the rest of the system. For example, asking construction teams to work overtime to make up for delays might speed up the rate of construction, but might also lead to more mistakes and therefore rework which could cause further delays.

In medicine, a patient's mindset can significantly affect their health. On a construction site, there is a strong guiding philosophy of health and safety that influences the decisions that are made within the system.

However, anyone who has worked on a construction project will be aware that the project is also impacted by external factors — things beyond the defined site boundary. For example: cost of materials, which are affected by global markets; supply chains and delays to materials, which can be altered by a ship crashing halfway round the world[2.3]; or changes to legislation that impact on the way we build. So let's consider a new boundary. What happens when we extend the system boundary beyond the edge of the construction site?

2.1.2 From project-scale to industry-scale

An important aspect in understanding systems theory is to appreciate how one system can be a subset or component of a much wider system.

In medicine, we can step back from an individual patient, themselves a system, and consider the health of whole groups of people, a much larger system. At this scale, new components come into play, such as hospitals and pharmacies. How these new components are connected to one another is the system structure. Prescriptions are one way that information is transmitted around this system; another is via public health campaigns.

At this scale, wider patterns of population health can be studied, while at the same time keeping in mind that a change to one part of the system will affect the rest of the system in some way.

Similarly, we can step back from considering an individual project to looking at all the projects in a town, region, or even a country. We can think of the components of the system as being the construction sites themselves, along with new components, such as the places where our materials come from. The organisation of the supply chain and the labour force are parts of the system structure. At this scale, wider patterns within the construction industry can be studied, such as trends in material cost or the impacts of material extraction.

In Table 2.1 we expand this idea. In the left-hand column we introduce different systems characteristics, with components at the bottom, moving right up to the system paradigm at the top. These are terms for characterising systems that we will be using throughout the book. In the right-hand column are examples of each of these system characteristics in the construction industry system.

We have introduced systems language to enable us to interrogate further sustainable and regenerative thinking. Having introduced terms for describing different characteristics of a system, we can simplify things a bit.

Table 2.1: System characteristics and examples of each from the construction industry

System characteristic	Construction industry examples
System paradigm — its guiding philosophy	The economic philosophy of continuous growth
System goal — the purpose of the system	To construct the built environment in a safe and profitable manor
Capacity to self-organise — how new ideas propagate within the system	Researching, testing and rolling-out of innovative construction processes, e.g., the development and implementation of: Building Information Modelling (BIM) technology; BREEAM; CLT
Rules — governing how the different components of the system can interact	Professional qualification of engineers; building regulations codes of practice; Heath and Safety at Work Act[2.4]
Feedback loops — mechanisms for reinforcing helpful processes and reducing unhelpful ones	Price of materials, reporting of profits to shareholders, reporting of BREEAM points, industry-wide health-and-safety data
Stocks — the availability of a variable	Availability of construction materials; plant; size of available workforce
System structure — how the elements of the system relate to one another	Supply chain connecting the construction site to the original source of materials, including processing plants; structure of the design and delivery team
Components — the individual elements	Individual construction sites, material mines, material processing plants

2.1.3 Introducing the Systems Bookcase

We find it helpful to think of Table 2.1 as a bookcase. On the shelves of the bookcase go books that describe different characteristics of the system (Figure 2.2).

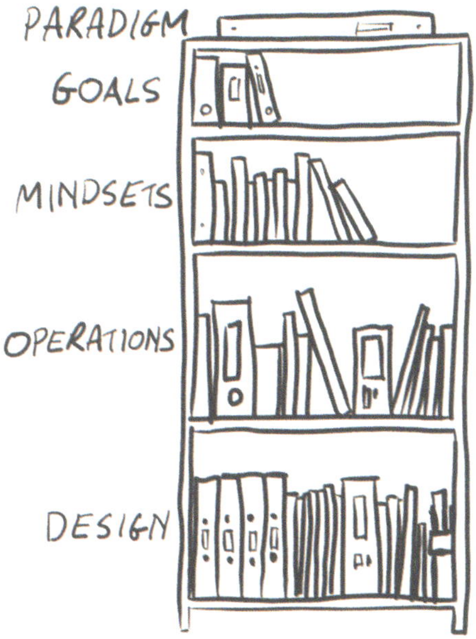

Figure 2.2: The Systems Bookcase

Each shelf has a different category:

- The top shelf is labelled **paradigm** and is for anything that describes the overarching socio-economic philosophy within which the construction industry (and all industry) operates
- The next shelf down is labelled **goals**. On this shelf goes a description of the goals that the system is trying to reach
- The shelf below that is labelled **mindsets**. On this shelf goes everything to do with how ideas emerge, the values that guide how the system operates
- The fourth shelf is labelled **operations**. On this shelf goes anything to do with how the industry is organised, from supply chains to design team structures. It includes how information is passed around the system and feedback systems. It also includes the rules and regulations (including codes of practice) that define how we operate

- The bottom shelf is labelled **design**. On this shelf go descriptions of the things we actually design and build, and descriptions of the resources we use to build these things

Now, as well as the system of categorisation, there is one additional, important organising principle for the books on each shelf: a book can only go on a shelf if it doesn't directly contradict the principles of the books on the shelf above. So, we are only able to include a particular way of designing on the bottom shelf if it is in line with the rules on the shelf above. In turn, on the second shelf up, we can only have rules that are in line with the mindsets described on the next shelf up. In turn, the only mindsets permitted are those that align with goals on the penultimate shelf. Finally, the only goals are those that are in line with the paradigm on the top shelf. Putting a book on the bookcase is a metaphor for saying whether something is allowed in this system.

To demonstrate, let's consider the example of using an innovative material on a particular project. The only way we can put this project on the bookcase, including its new material, is if this material is compatible with the existing design codes on the shelf above. If the rules don't allow the material, then that project is not allowed on the bottom shelf. That is, unless the culture and values higher up in the bookcase allow the rules to be changed to accommodate the new material. We do, of course, allow some degree of material innovation, but not everything. What is allowed is what is governed by the system goal: that the material must be safe and the project must be profitable.

For example, the regulatory approach between the US and Europe (and in this case the UK) is quite different. The Eurocodes are predicated on EN 1990[2.5]. These rules include, if you look for them, information on how to design with materials not directly covered by the code, through testing (have a look in Annex C and D). This is how CLT, a non-codified form of timber construction, has been constructed with in Europe for years. In the US, talking to engineers who work in that environment, a similar code does not exist, making it much more challenging to build outside the codified guidelines. The exception being in California where specific seismic codes apply. Of course, if you have ever tried to design a building from a non-standard material, you will know that the limitations are not just around codes, you still need to persuade insurers, financers and building control.

This system hierarchy for organising the books on the bookcase reflects a simplified model of Meadows' 12 Levers of Systems Change (Appendix A).

Now that we have established the metaphor of a Systems Bookcase, we will use it to compare sustainable and regenerative thinking.

2.2 SUSTAINABILITY

Consider the resources we often take for granted: clean air, clean water, fertile soil, stable climate and biodiversity. These are our life-support systems, critical to human existence[2.6].

These systems rely on living webs of unfathomable complexity, which help to keep them in balance, provide resilience and enable them to evolve as conditions change. If we viewed society as a business, and these life-support systems were its supply chains, wouldn't we devote considerable resources to ensuring those systems not only worked well but got progressively stronger over time?

Regrettably, the reality is that humans, particularly in the Global North, are doing the opposite. It is important to recognise that engineers generally act with good intentions and have created enormous improvements in human health and wellbeing worldwide. But, as we are only now starting to realise, if we were to sum up all our collective design actions, the net impact of all these actions on our living systems is negative. Air and water polluted, soil degraded, significant reductions in biodiversity and an increase in harmful greenhouse gases in the atmosphere. And it is not just the health of these individual systems but also their resilience and even their capacity to evolve in response to this change that we have damaged.

A hard-hitting description of humankind's destruction comes from *Where the Leaves Fall*[2.7], a magazine that explores human connection with the living world:

> "This incessant violence against the landscape has a very violent dynamic. It doesn't just happen here in Brazil. Engineering is failing across the planet. It focuses on natural resources, on water, on the landscape, on the mountains — it flattens everything."

In the introduction we present the goal of regenerative design as being "for human and living systems to survive, thrive and co-evolve" (Appendix B contains some standard definitions of 'regenerative design'). But what if the opposite is occurring? What if the sum of our actions is that biodiversity is in decline, that soil health is reducing, that air is becoming more polluted and temperatures are rising. Sadly, we know that this is the case. So, if we are not currently working in a regenerative way, we need to accept that we are in fact working in a **degenerative** way.

So how have we ended up in this situation? We consider four factors.

2.2.1 Sustainability in the current paradigm

Our first consideration is to look at the place of sustainability within the current paradigm in the construction industry. The idea of sustainability is old enough that many reading this book will have been working with sustainability for most, if not all, of their careers. In 1987 the United Nations Bruntland Commission defined sustainability as:

> "Meeting the needs of the present without compromising the ability of future generations to meet their own needs".

So, where does the concept of sustainability fit on the Systems Bookcase (Figure 2.3)? If sustainability were truly our overarching way of understanding the world, then it would sit near the top — serving as a goal that governs all aspects of the construction industry.

Figure 2.3: Where does sustainability fit on the Systems Bookcase?

However, the reality is that in most cases, sustainable considerations have been largely limited to a narrow view of the procurement of materials and the impacts of a site on its local community and ecosystem. In this light, it is often treated as a tick-box exercise on the operations shelf, or even as a design consideration on the bottom shelf.

Sustainability as marketing

Consider the following scenario as an illustration of how sustainability can be subverted to a lower-level systems intervention.

A senior engineer we interviewed for this book once had a meeting with a large asset owner to advise them on a sustainability strategy for their estate. The client's Head of Sustainability suggested involving their superior in the next meeting. Anticipating a chance to make a significant impact, the engineer was thrilled, assuming that the Head of Sustainability would report directly to the board or the CEO.

To the engineers' disappointment, they discovered in the next meeting that the Head of Sustainability was, in fact, under the jurisdiction of the Head of Marketing. This is indicative of the fact that sustainability, instead of being an intrinsic part of the organisation's strategic planning and decision-making process, was side-lined as an aspect of their public relations strategy.

2.2.1.1 The bottom line

All engineering practitioners will be aware of the importance of the bottom line. The need for all projects to make a profit at all levels. The reality is that this isn't always the case of course — you win some and lose some — but if your winnings outweigh your losses then all is well. You have made a net gain.

Sustainability has, at its heart, a desire for outcomes that consider more fully the impacts of projects. This philosophy is captured in the notion of the 'triple bottom line' — the need to find the best outcome across financial, social and environmental requirements. However, reflecting on his industry experience, James found that value was primarily evaluated in financial terms. The prevailing practice was to focus primarily on cost — and only if the cost were comparable (or lower), would sustainability factors be considered.

James uses a footballing analogy to describe this phenomenon. Economics is the primary scoring mechanism (three points for a win, one for a draw, none for a loss), while social and environmental considerations are the equivalent of measuring goal difference. Only when the economic scores are the same do we take into account the 'goal difference' of the social and environmental impacts.

Which raises the question 'why?' Why did we measure everything in monetary terms, instead of in other terms. And why did we (and everyone in the system) need to make a net gain year on year?

The answer helps us to understand the overarching philosophy of the system — the economic paradigm of continuous growth. It is a philosophy so entrenched in our thinking, imagining a different way of doing things is challenging.

Let's be clear, not everyone working in industry agrees with, or supports this system. We may advocate for different approaches, we may suggest that surely we should value our ecosystem more highly than our financial system. But coming back to the Systems Bookcase, where the boundary is the construction industry, the paradigm for our industry is continuous growth. Without challenging the paradigm, sustainability can only operate at the lower levels of the Systems Bookcase.

2.2.2 Carbon reduction: a case in point

Our second consideration is to look at how we have approached carbon reduction. Around the turn of the century, the building industry began responding to concerns about the impacts of construction by talking about adopting the principle of sustainability. At best, sustainable design seeks to do no harm — make things no worse than when you found them. But this approach says nothing about actually making things better than you found them; of fixing the damage already done.

While many people have been pioneering principles of sustainable design in their work, very few projects are fully sustainable — very few actually do no harm. And so, most projects are still doing damage to the life-sustaining systems that we rely on. The result has been to aim low and miss. We aim to do no harm but end up doing harm.

The 2018 IPCC report[2.8] raised public awareness about the severity of the climate crisis. Within a year the UK Government had set a target of reaching net zero carbon emissions by 2050. Many in our sector are now focusing on how to decarbonise structural engineering. Yet, as with sustainability, the problem we have is that the target we set is one of doing no harm — stopping putting carbon dioxide and other greenhouse gases into the atmosphere. It says nothing about repairing the damage done — in this case, sequestering the carbon dioxide already in the atmosphere, which even if we stopped emitting carbon dioxide today, could still be enough to trigger irreversible changes in our climate. While eliminating carbon emissions from the built environment sector is a necessary and important step, the net zero target is not enough on its own. And on our current trajectory, we are less and less likely to achieve it.

This restriction of sustainability to lower-level systems interventions allows for superficially sustainable measures that are permitted because they do not significantly challenge the paradigm of continuous growth for the sector. Recall from Section 2.1.3 that a book is only allowed on a shelf if it doesn't significantly contradict what is written on the shelves higher up. When sustainability interventions are restricted to the lower shelves of the bookcase, they are sufficiently low down so as not to significantly challenge the paradigm, and so they are tolerated without meaningfully changing how the system operates.

From this examination of sustainability and carbon targets we see two recurring themes:

1. Aiming low: setting a target that merely aims to prevent further damage without addressing the need for repair.
2. Missing the target: failing to achieve even these modest goals because we are limited by a higher need on the Systems Bookcase; the need for profitability.

These patterns (Figure 2.4) underline the need for a new approach, one that goes beyond setting low targets and establishes a new goal of **repair**.

Figure 2.4: Our current approach is to aim low (do no harm) and to miss. What would happen if we could see beyond doing no harm?

2.2.3 The problem of focusing solely on carbon

Our third consideration is the problem of focusing solely on carbon. In his book *Designing Regenerative Cultures*[2.9], Wahl asks the pertinent questions:

> "Have we defined our goals correctly? Are we trying to maximise isolated parameters or to optimize the whole system?"

We are currently facing up to the realities of the climate crisis and focusing on reducing carbon. Don't get us wrong, this is a good thing! We have been arguing for this approach since the 2000s (and others since the 1970s). We celebrate the recent work of the IStructE in this space and the continual flow of good quality guidance and leadership. But we also need to avoid the mistake of thinking that it is the only issue we need to consider.

Reducing greenhouse gases to preindustrial levels would have the benefit of stabilising the climate, but on its own would do little to repair the other living systems that are also being destroyed. If we banned concrete tomorrow and replaced it with timber, the impact on our global forests would be immense.

When we start to think about all the elements of our ecosystems that need urgent attention, we realise we can't design for any one of these metrics, we must design for the health of the system as a whole. While eliminating carbon emissions is an essential part of what we need to do, the higher-level goal needs to be much wider than just focusing on carbon: it needs to focus on the health of the whole living system, humans included.

So far we have considered sustainability within the paradigm of growth, the impact of aiming low in our targets and missing, and the problem of focusing solely on carbon. But there is one more piece of the jigsaw we need to consider.

2.2.4 The missing piece: healing

Our fourth consideration is the question of healing. Returning a system to health isn't as simple as hitting reverse. When we deforest a landscape, we can quickly create a desert. Without tree roots, the soil is easily eroded by wind and rain. Without the soil, the complex microbial systems in the ground disappear. But the process is not simple to reverse. Trees planted in poor soil can't become the towering size of before. The system complexity needs to redevelop before the system will return to its original richness (Figure 2.5).

Figure 2.5: Oliver recently saw the living world healing itself on a visit to Mount Etna. Lava flows destroy everything in their path. But once the lava has cooled, lichen forms on the lava, helping to retain water near the surface. Plants can then grow and before long, trees start to re-establish themselves. Pioneer species such as birch are fast-growing, creating leaf litter that starts to build up the soil that other species of tree can grow in

Let's recap this chapter so far. In Section 2.1, we introduced systems thinking to view the construction industry through a new lens. We've used the Systems Bookcase model to create a system hierarchy in construction; from the physical assets that we build at the bottom, up to the overarching paradigm at the top. The model helps us see how different characteristics of the construction industry system that we work in, exist in a hierarchy and that this hierarchy exerts control over what is and isn't allowed.

In Section 2.2, we looked at the impacts of the construction industry on the ecosystems that support life and saw that despite good intentions of being sustainable, our industry continues to do harm. We used the Systems Bookcase to reveal the limits so far to sustainability. This model reveals how sustainability, despite its potential to be a governing paradigm, is largely relegated to a lower shelf in our system. This superficial integration of sustainability has allowed the industry to continue with good intention, without addressing the deeper systemic issues at play.

We saw how setting a target to do no harm and then continuing to miss this target leads to continued harm. We also considered the limitations focusing on one metric (carbon) as a measure of our environmental impact. Finally, we considered what we see as the missing piece of the jigsaw: how to help living systems heal.

The stage is now set for introducing regenerative design, set within the context of holistic thinking.

2.3 A NEW HOLISTIC APPROACH

One of the biggest challenges with our current economic models is that the boundaries are drawn so that the limits of our ecosystem are outside the system.

However, within economics there has been a movement to abandon the pursuit of limitless growth of GDP[2.10]. This school of thought recognises that our economy must operate within the boundaries of our ecosystem's limits. One manifestation of this principle is the concept of 'doughnut economics' (Figure 2.6).

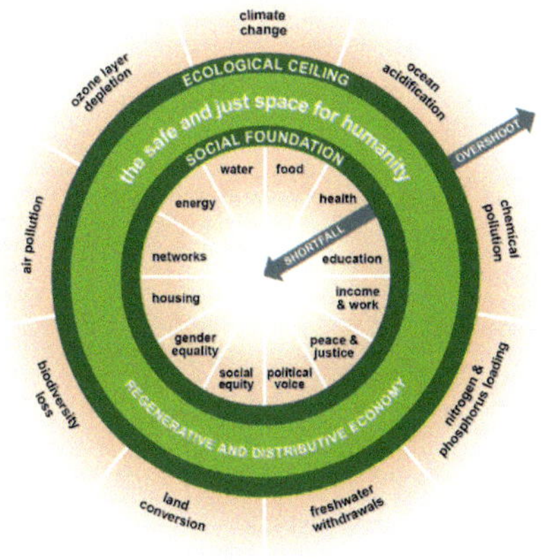

Figure 2.6: The doughnut of social and planetary boundaries

Within this framing, the economy should exist between a lower threshold of providing for humanity's needs, without exceeding the upper limits of what the ecosystem can withstand. This progressive approach is now gaining traction at city-scale (e.g., in Amsterdam[2.11]).

To shift from a perspective of continuous growth, to growth that exists up to a ceiling is a fundamental change of philosophy. When we acknowledge that we can't just fix things by growing more, we must rethink how we can use what we have to create the society we want. This isn't to say that all growth is bad, but we can't assume all parts of the system can continue to grow forever. Some economies need to grow in order to create a basic social foundation for a healthy life. But at some point, we will reach the limit of what our planet can sustain, and as the climate and biodiversity emergency is showing, in many places we have exceeded these limits already.

This change represents a shift from a continuous growth paradigm to one of holism. According to the dictionary, holism is the theory that 'parts of a whole are in intimate interconnection, such that they cannot exist independently of the whole, or cannot be understood without reference to the whole, which is thus regarded as greater than the sum of its parts'. Although the words 'holism' and 'holistic' were defined in the 1920s, it is an ancient approach reflected in many traditions from Indigenous practices to the Chinese Tao[2.12].

At first, this discussion of holism may seem unrelated to structural engineering. However, as the Systems Bookcase shows, the paradigm we are working to is the most powerful influencer of what we are able to design on the ground. Recognising that we need to live within our ecosystem limits, acknowledging that we are reliant on systems of unfathomable complexity, and understanding that we need to restore these to health, all point towards the necessity of a new paradigm — a paradigm of holism.

2.3.1 Regenerative design

As our industry, including clients and government, responds to the climate and biodiversity emergency, a new, more holistic approach to design and construction is starting to gain traction — regenerative design.

The Engineers Declare a Climate and Biodiversity Emergency statements of Canada, France, the UK and Singapore, signed by more than 450 firms globally, specifically mention regenerative design. The wording in the UK structural engineers' statement, signed by 270 practices at the time of writing, is as follows:

"We will seek to:

- Advocate for faster change in our industry towards regenerative design practices and a higher Governmental funding priority to support this
- Adopt more regenerative design principles in practice, with the aim of providing structural engineering design that achieves the standard of net zero carbon"

This concept of acting regeneratively, while already established in sectors such as farming and healthcare, is still relatively new in our sector.

In its early stages, regenerative design has been associated with a broad range of ideas — from biomimicry to circular design, from community resilience to rewilding. But we find it most helpful to think of regenerative design as a systems goal. In simple terms:

> 'The goal of regenerative design is for human and living systems to survive, thrive and co-evolve'.

Using the Systems Bookcase concept we introduced earlier, as a system goal, regenerative design goes high up on the bookcase, consistent with the top shelf book of holism (Figure 2.7). This definition seeks to underscore the importance of integrating our built environment, including the whole system of construction, harmoniously with living systems. Integration is consistent with the interconnection inherent in holistic thinking.

Regenerative design transcends the notion of sustainability, envisioning a future where integrated human and living systems are not only repaired, but have recovered their capacity for self-repair.

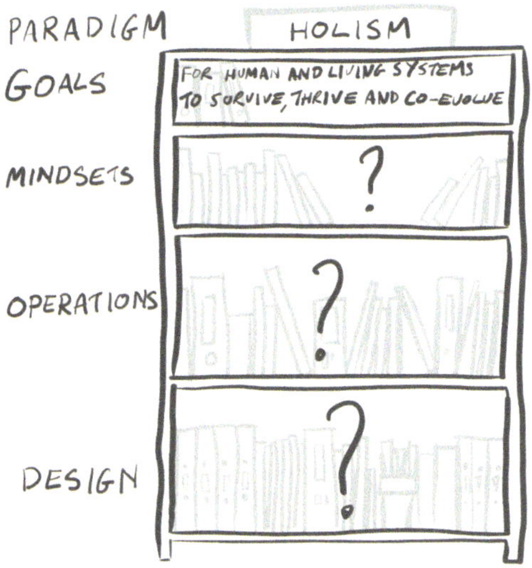

Figure 2.7: Setting the goal on the Systems Bookcase

This goal of surviving, thriving and co-evolving is a powerful statement of intent for design of the built environment. If we can learn how to design regeneratively, then when we design buildings, towns and cities we have the opportunity to not only minimise harm to the environment but actively contribute to its healing and in doing so, secure a better life for ourselves and future generations.

2.4 CONCLUSION

At the start of this chapter, we asked the question: Why do we need to think regeneratively?

To help answer this we used systems thinking to help us see the construction industry as a system. From this systems perspective, we described how everything that happens in construction is subject to rules, regulations and goals, which are all ultimately in line with an overarching paradigm in the system. We introduced the Systems Bookcase model as a simplified way of understanding how different characteristics of a system control what does and doesn't happen in that system.

We then went on to look at the impacts of the construction industry on our ecosystems — despite the industry trying to become more sustainable for more than 20 years. Using our systems framing we saw that the impact of sustainability has been limited by: the paradigm of continuous growth; the problem of setting a target of doing no harm and missing that target; the issue of mainly focusing on carbon; and the failure to consider how we heal the damage already done to our planetary life-support systems.

Together, these factors point to the urgent need for a new way to frame how we design so that the impact of our work is an increase in the health of the planet we live on. We saw that the paradigm of holism, one that respects planetary boundaries and the need for all that live on that planet to thrive, is gaining traction as a philosophy with which to respond to the climate and biodiversity emergency.

It is within this context that regenerative design has emerged. The aim of regenerative design is for human and living systems to survive, thrive and co-evolve. As a systems goal, regenerative thinking sits high on the Systems Bookcase, and therefore has the potential to completely reshape how the built environment industry operates. Framed within a holistic paradigm, it more clearly articulates our interconnection with living systems and the need for all living things to thrive in balance.

Shifting from thinking sustainably to thinking regeneratively invites us to consider how we challenge the whole paradigm within which we are working so that our industry can be an overall force for good. Given the scale of the negative impact of our sector, this shift in our thinking couldn't be more urgent.

2.4.1 What's this got to do with me?

At this point we anticipate that you might be wondering about the practical implications of regenerative design and the role of structural engineers in this concept. You may well be thinking that while this is interesting, what does it have to do with you? That engineers are not able to change the paradigm, so must continue to do their best within the constraints that we have. You may find yourself thinking, this is all very well, but isn't it up to the client, or the architect, or some other decision-making body to bring about this change?

To help us answer these questions, in the next chapter we introduce the Three Horizons model. This model helps us see how one dominant paradigm gives way to the next, and how structural engineers have an important role to play in creating a transition towards a regenerative built environment sector.

CASE STUDY 1

—

Throughout this book we will include case studies. However, this first case study is not really a case study. It is an 'anti case study'.

It is all the projects that didn't consider sustainability.

It is all the projects where decisions were made before we even got involved, which have led to a less sustainable outcome — from longer than necessary spans to existing buildings being demolished.

It is all the projects where sustainability was discussed but removed due to cost considerations.

It is all the projects where sustainability was included in the design, but value-engineered out.

It is all the projects where BREEAM points were selected not due to positive impact but the lowest financial cost to achieve the minimum required outcome.

It is all the projects where information that could help others to be more sustainable was kept for competitive advantage rather than shared to maximise impact.

It is all the projects where the impact is not regenerative, but degenerative.

It is the vast majority of projects we have ever worked on.

Feel free to spend a moment thinking about which projects you would like to add to this space.

REFERENCES

2.1 Meadows DH. *Thinking in Systems.* London: Chelsea Green Publishing, 2015.

2.2 Senge P. *The Fifth Discipline: The Art and Practice of the Learning Organization.* New York: Currency Press, 1990.

2.3 The Guardian. *How a container ship blocked the Suez canal — visual guide.* Available at: https://www.theguardian.com/world/2021/mar/24/how-a-container-ship-blocked-the-suez-canal-visual-guide [Accessed: December 2023].

2.4 Health and Safety Executive. *Health and Safety at Work etc Act 1974.* Available at: www.hse.gov.uk/legislation/hswa.htm [Accessed: December 2023].

2.5 European Commission. *EN 1990. Eurocode. Basis of structural design.* 2002. Available at: https://eurocodes.jrc.ec.europa.eu/EN-Eurocodes/eurocode-basis-structural-design [Accessed: December 2023].

2.6 Wallace-Wells D. *The Uninhabitable Earth: A Story of the Future.* London: Allen Lane (Penguin Random House), 2019.

2.7 Where the Leaves Fall. Issue 8. *Crisis/Resilience/Recovery.* Available at: https://wheretheleavesfall.com/shop/catalogue/wtlf-issue-8_46/ [Accessed: December 2023].

2.8 Intergovernmental Panel on Climate Change. *IPCC Special Report: Global Warming of 1.5°C.* Available at: www.ipcc.ch/sr15 [Accessed: December 2023].

2.9 Wahl DC. *Designing Regenerative Cultures.* Bridport: Triarchy Press, 2016.

2.10 Jackson T. *Prosperity without Growth: Economics for a Finite Planet.* Abingdon: Routledge, 2011.

2.11 Doughnut Economics Action Lab. *Amsterdam City Doughnut.* Available at: https://doughnuteconomics.org/stories/1 [Accessed: December 2023].

2.12 Fleming D. *Lean Logic.* Available at: https://leanlogic.online/[Accessed: December 2023].

CHAPTER 3

Foundations of regenerative thinking

In the previous chapter, we made the case for transcending sustainability to a new, regenerative approach to design, grounded in a holistic philosophy. In this chapter we ask: **How can we develop regenerative design within a paradigm of continuous growth?**

To begin, we introduce the 'Three Horizons' model, a tool for envisioning how one paradigm gives way to another (Section 3.1.1). This model, when coupled with our Systems Bookcase, illustrates the potential transition from our current practices to a regenerative future.

Then, to provide a more tangible picture of regenerative design, we turn to the best regenerative example we know — living systems themselves. We briefly examine their characteristics at a systems level, seeking to understand what makes them survive, thrive and evolve. These living systems can serve as our blueprint for regenerative design.

Finally, we turn our attention to the role of the structural engineer in this industry-wide transformation.

The foundations we establish in this chapter pave the way for the practical suggestions we make for regenerative practice in Part 2 at the level of design, operations and mindsets.

3.1 MAPPING THE PATH TO REGENERATIVE

If you are finding it difficult to imagine what a regenerative construction industry looks like you are not alone, given how starkly this idea contrasts with the industry we see around us. So how do we reconcile our ambition for a different way to design with the reality of day-to-day work? And how can we envisage our current ways of working giving way to a regenerative approach? To help answer these questions we will consider the Three Horizons model.

3.1.1 Introducing Three Horizons

The Three Horizons model, developed by Bill Sharpe, is a widely used model for futures thinking[3.1]. It is useful for helping us understand why future scenarios might seem elusive, and gives us a strategy for helping us transition from an existing state to a preferred one.

The model defines three distinct horizons:

- Horizon One (H1) represents our current worldview
- Horizon Three (H3) envisages the world we want to build
- Horizon Two (H2) enables the transition from H1 to H3

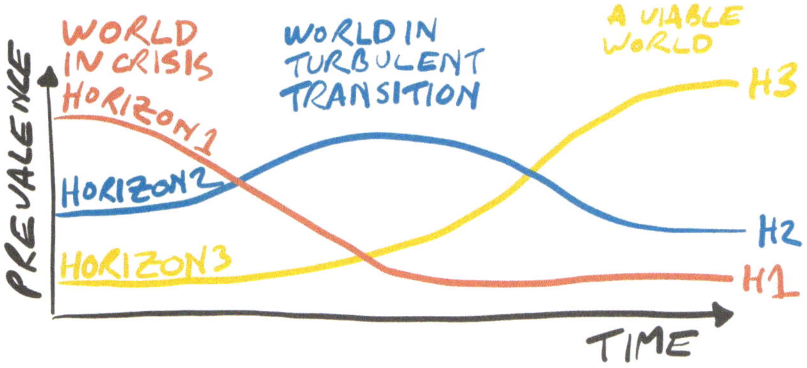

Figure 3.1: The Three Horizons model

3.1.2 A simple construction industry example using the Three Horizons model

Let's introduce the Three Horizons model by looking at how a safe working culture has been introduced into the construction industry. To begin we'll rewind the clock to the 1980s/90s, when the UK construction industry was a high-risk work environment with a large number of accidents and frequent fatalities. This scenario represented Horizon One practice; the existing situation, where the prevailing ways of doing things were so entrenched that doing things any other way seemed impossible.

And yet pressure for change was inevitable — the unacceptable human cost, changing ethical standards and the financial impacts of accidents meant that that way of working could not be sustained forever. Change was coming, but in what form would not have been clear.

This initial phase is a critical state in the Three Horizons model; the time at which the present model is unstable but the future remains uncertain.

Now, at that time, a radical alternative future would have been to imagine a workplace in which going to work actually made us healthier. This vision represents Horizon Three in the model — something that would have seemed like a radical departure from the status quo.

The subsequent implementation of health and safety regulations represents Horizon Two — an event, technology or approach that enables a transition from the current state to the desired future state. It doesn't fully realise the future we envisage, but it is a stepping stone towards that goal. What is interesting about Horizon Two interventions is that they can go one of two ways. They can indeed usher in Horizon Three, in which case they are referred to as 'H2+' interventions. In the case of health and safety regulation, the construction industry has certainly become safer, in which case health and safety regulation could be regarded as a H2+ intervention. But if doing risk assessments becomes a checkbox exercise that didn't actually change behaviour, then rather than drive the system to a new way of working, these regulations are merely extending the life of the current way of working. In which case the intervention is known as 'H2-'.

The implementation of H2 technologies or policies is another crucial state in the Three Horizons model because while they can usher in a new future, they can be subverted to prolong the life of the existing system.

An interesting feature of the model is that while we can think of H3 as the dominant future scenario, it actually begins, at a small scale, in the present. In the metaphor of horizons it is the tiny sliver visible in the far distance that you can see from your current location (Figure 3.2).

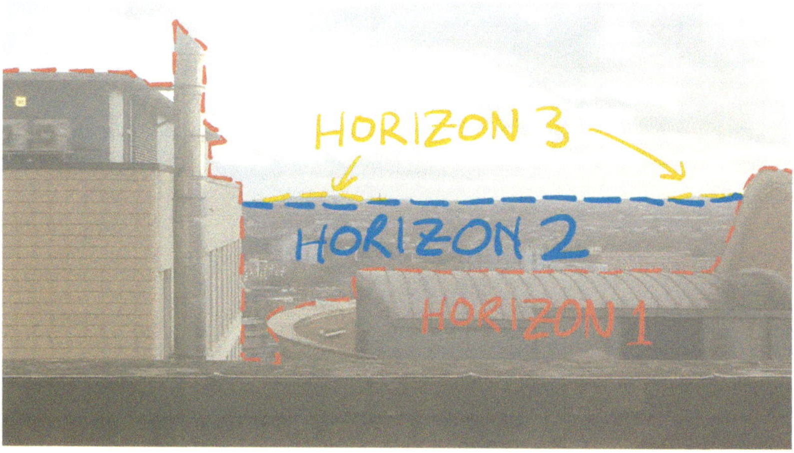

Figure 3.2: How the Three Horizons appear from our current context

In our scenario, an example of H3 in the present, is a construction company that is pioneering safer ways of working ahead of what any legislation might require. Pioneers of future-thinking ways of operating pave the way for legislative changes because policymakers often want to know what might be possible before it gets set in legislation.

Identifying H3 in the present is a critical step in a Three Horizons analysis because it helps us recognise pioneering work that we should protect and grow in order to usher in the future that we want to build.

In this section we have used the transition to safer ways of working in the construction industry to illustrate how the Three Horizons model works. Now let's use the model to show the transition of the construction industry away from current practice towards more viable, regenerative working.

3.1.3 From the present to the future

In Chapter 2 we introduced the idea of the Systems Bookcase to illustrate how the various components of the construction industry must adhere to the governing rules and overall paradigm of the system. By plugging this Systems Bookcase model into the Three Horizons framework (Figure 3.3), we can create the transition from our current paradigm to a

Figure 3.3: Combining the Three Horizons and the Systems Bookcase models

holistic one where the construction industry operates regeneratively. Let's take a closer look.

3.1.3.1 The bookcase of the present

The bookcase on the left in Fig. 3.3 represents the construction industry in the present. This is the model we introduced in Chapter 2, but now with added colour.

Most of the books are red, consistent with the dominant philosophy of continuous growth — these are the Horizon One books. However, there are a few blue books representing changes to the current system that are a stepping stone to a different paradigm — Horizon Two books. While these don't disrupt the system on a small scale, their growth could shift the paradigm.

There are also a handful of yellow books, rare examples of Horizon Three in the present — regenerative ways of working that already exist. There are only a few examples of these because they are largely incompatible with the current paradigm, and where they do exist, they can often seem very theoretical, as examples are difficult to identify.

The bookcase of the present highlights why it is currently challenging to act regeneratively. Regenerative principles are so divergent from the prevailing paradigm that they are seldom allowed in practice. **If you're struggling to see how regenerative design applies to your work as a structural engineer, this resonates with the significant shift that is needed from our growth-orientated economy.**

However, the world is in crisis. We're seeing escalating impacts of climate breakdown, the unravelling of the global extraction-based supply chain, and downward trends in human health indicators. These are the hallmarks of the inevitable downfall of an economy built on the unsustainable premise of continuous growth. Horizon

One is starting to fall, creating space for different books on the bookcase.

3.1.3.2 The transition bookcase: the stepping stone

As the climate and biodiversity emergency intensifies, we transition from the status quo to a transition state, represented in the next bookcase in Fig. 3.3. In this bookcase, the paradigm has moved away from continuous growth but hasn't fully embraced holism. The blue books represent technologies, initiatives and approaches consistent with this transition paradigm, which could potentially lead to a regenerative future. However, it's unclear whether they represent a genuine stepping stone towards a regenerative future (H2+) or merely prolong the current system (H2-).

Net zero targets, circular design and artificial intelligence all represent possible H2 technologies and approaches. Let's briefly look at each of these.

Net zero targets and legislation sits on the middle shelf of the transition state bookcase; the shelf for operations. Net zero approaches to design are increasingly going to influence the design of the projects on the bottom shelf. But on their own, are they enough to develop a mindset on the next shelf up that sees all our ecological systems as connected? Or are net zero targets likely to drive the exploitation of other ecological systems in pursuit of eliminating carbon?

Circular design is another approach that sits prominently on the middle shelf of the transition bookcase. Circular design guidance will encourage the reuse of existing materials on existing projects and design for disassembly and reuse. But will circular design act as a stepping stone to say, the creation of generative material supply chains that increase ecosystem health the more material they produce? Or will circular design provide an excuse for continuing to extract materials as long as they can be reused? And how do cities and communities in the Global South adapt, when they don't have the material stores of existing building that we have built up over the last few centuries? Will circularity enable them to develop in a fair and just way?

And sitting prominently on the middle shelf of the transition bookcase is all the ways that **artificial intelligence** is changing the way the construction industry will operate. But will the power of this emerging technology enable us as designers to better tune in to the needs of specific human and living systems, finding the best possible designs to enable these systems to survive, thrive and co-evolve? Or will artificial intelligence further separate us from the needs of these systems and accelerate their exploitation — replicating and reinforcing our current ways of designing.

This analysis of H2 technologies helps us to understand what the blue books on the bookcase of the present are. These are the ways in which the current system is already adopting net zero approaches, circular design and AI. But as we saw earlier, while H2 technologies usher in a new future, they can also be subverted to prolong the life of the existing system.

3.1.3.3 The bookcase of the future: for the books we need to write

The bookcase on the right of Fig. 3.3 represents the regenerative construction industry of the future. At the top is the H3 paradigm of holism, and beneath that the goal of regenerative design.

Most of the books are yellow, reflecting this new paradigm. Some blue books from the transition period remain, and a few red books remain — those practices from the current system that are still needed and allowed in the future.

While the holistic paradigm and goals of regenerative design are firmly established, we don't yet know the content of the yellow books. In other words, we're clear about our goal of regenerative design, but the specifics of how this translates into each level of the system (design, operations and mindsets) are still unknown.

We can just make out the highest shelves of the regenerative system, but what it looks like in detail is obscured by our current world view. **In fact, we see the biggest barrier to transitioning to a regenerative future as the inability to imagine what this future looks like.**

But if we can start to define what a regenerative future looks like, we can identify current instances where we're already adopting these approaches (the yellow books on the bookcase of the present) and the transition approaches that could lead to a regenerative future (the blue books on the bookcase of the present).

We can apply this analysis at each level of the bookcase: at the level of design; at the level of operations; and at the level of mindsets. At every level of the construction industry system, we need to identify the H2 and H3 approaches that we must nurture, and the H1 approaches that we need to phase out. To apply this analysis, we need a clear guide to what regenerative practice means at all levels of the system. What better guide do we have than the oldest regenerative example on the planet — living systems themselves?

3.1.4 Conclusion: mapping the path to regenerative

In this chapter we are asking how can we develop regenerative design within a paradigm of continuous growth? The first part of the answer comes from the Three Horizons model. The model shows how a system in crisis can transition to a more viable way of working. Critical to that transition is choosing the right transition steps so that we are genuinely creating stepping stones to that viable future, rather than merely prolonging the current state.

Key to making the right transition is having a clear idea of where we are trying to get to.

The next part in answering the question of how we develop regenerative design is to learn from living systems.

3.2 LEARNING FROM LIVING SYSTEMS

Learning from the living world is nothing new; structural engineers have been doing this for years. But just as through biomimicry we seek to learn from the incredible structures that living systems build, in regenerative design we can learn from how ecosystems work at a systems level to provide the conditions for thriving within the limits of what the ecosystem can sustain.

A living world example that we can learn from and that seems to resonate well with the engineers we speak to is that of the beaver. In Europe the beaver was hunted almost to extinction. But where it is re-introduced, the impact on the local ecosystem is profound. Through the creation of beaver dams (Figure 3.4), water is held in the landscape for longer, increasing drought-resistance, increasing species diversity and boosting carbon sequestration through damper soils.

For a single species, beavers have a disproportionately positive impact on their ecosystem, so much so that in ecological terms they are known as a 'keystone species'.

You can also see from aerial photos where humans are active in the landscape. We too have a disproportionate impact on our ecosystem for a single species, but in the negative sense. Estimates are that fully one third of species on Earth have vanished due to human activity, and the rate continues[3.3] (Figure 3.5). In the Global North, where we have lost the ability to live in harmony within our ecosystem, we have

Figure 3.4: Beaver dam

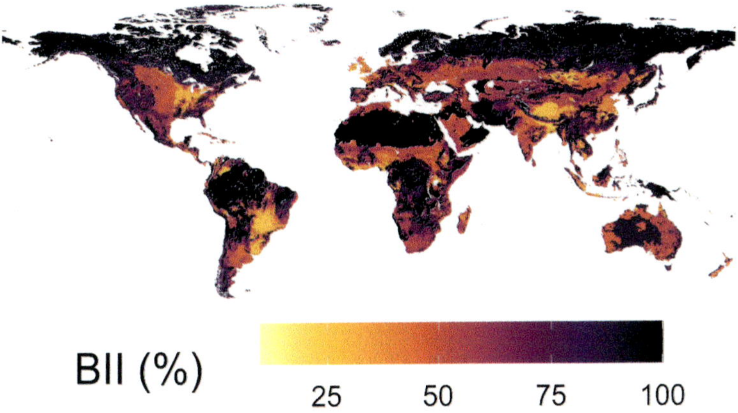

Figure 3.5: Loss in biodiversity across the globe

become the opposite of a keystone species. If we were to classify ourselves as a symbiont with the ecosystem, our sub-class would be that of parasite. Parasitism involves one organism living at the expense of another.

But what if we could change that narrative? What if we too could be a keystone species? That is the vision that regenerative design seeks to realise. The starting point is to use the living world as our guide. After all, nature has had billions of years of figuring out how to create systems that survive, thrive and co-evolve.

We have identified three characteristics of living systems that can help us understand how these systems can survive, thrive and co-evolve, help us envisage what a regenerative built environment might look like, and guide us on how we need to organise our industry to make that happen.

3.2.1 Connection

The first characteristic of living systems we'll explore is connection. Living systems are highly interconnected, facilitating communication and feedback from one part of the system to another. This interconnection is complex and forms the basis for the efficient functioning and resilience of these systems[3.4].

3.2.1.1 Feedback loops in a forest

In living systems, interconnection is exhibited in the way the elements of the system (the living organisms themselves and their environment) are interlinked. This connection gives rise to two types of feedback loop that govern system behaviour: balancing and reinforcing feedback loops. These loops are essential for maintaining the system's balance and resilience[3.5].

For example, in a forest, a balancing feedback loop occurs between deer and the saplings they feed on. When the deer population is high, they overgraze on saplings, reducing the number of saplings and their ability to grow and to reproduce. As the number of saplings reduces, so does the available food for the deer, and so the deer population decreases. The lower deer population enables the saplings to recover, which in turn allows the deer population to grow again. Balancing feedback loops help to maintain stability in a system, preventing any one population (or 'stock', in systems terms) from becoming too dominant.

Again, in a forest ecosystem, there is a reinforcing feedback loop that exists when trees return to an area that has been cleared (e.g., from felling, a storm or fire). As the pioneer species return (e.g., birch in the UK) their leaf-fall creates a richer, safer environment which in turn encourages more trees to grow, which will create even more leaf litter. And so on and so on. This is life reinforcing life (Figure 3.6).

These examples of forest feedback loops are highly simplified. The real interactions are much more complex and involve many more interactions. To delve deeper into the interconnected world of woodlands, refer to *The Hidden Life of Trees*[3.5].

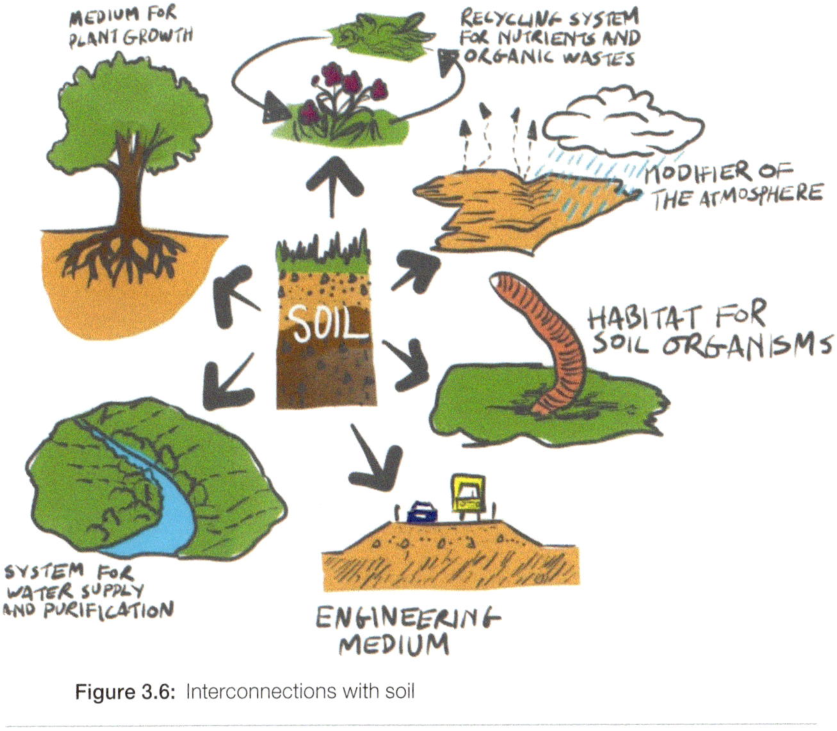

Figure 3.6: Interconnections with soil

3.2.1.2 Implications for the built environment

Living systems are stacked full of layers of balancing and reinforcing feedback loops. These overlapping layers enable feedback on feedback and redundancy that ensure system resilience, enabling living systems to survive and thrive.

Reflections on how we as humans were once more interconnected with living systems can offer insights into the potential shape of a regenerative construction industry. But we're not advocating a return to practices of the past. Instead, we're questioning: **What would it mean for us to reclaim that level of interconnection in our modern design and build processes?**

By weaving a higher level of interconnection into our design, operations and mindsets, we can enhance the self-regulation and resilience of human-living systems. But this idea of interconnection extends far beyond feedback loops. It embraces a holistic philosophy that acknowledges each part of the living world as a component of a complex whole that includes us.

Returning to our Systems Bookcase model, **we can use this understanding of interconnection to prioritise design approaches, operational methods, and mindsets that will support our transition to more regenerative ways of working.**

3.2.2 Supply

The second characteristic of living systems we'll explore revolves around the supply of resources. Living systems use the sun's abundant energy (either directly or indirectly) and renewable mineral resources[3.4]. All the complexity, diversity, vitality and evolutionary capacity of living systems is created from these self-renewing resources. To ensure that we too can survive and thrive within our ecosystem boundaries, we too must figure out how to participate in this web of renewal.

3.2.2.1 Symbiosis

Renewable systems are ones which continue to reproduce. A well-managed farm can produce crops annually while also enabling biodiversity to flourish, managing water and minimising impact downstream (this is beautifully illustrated in *English Pastoral* by James Rebanks[3.6]). The act of living together is called 'symbiosis' and provides the key to helping us understand how we can live with renewable systems. Symbiosis can be categorised into three relationships: 'mutualism', 'commensalism' and 'parasitism'[3.7].

Mutualism describes a situation where all parties benefit. When beavers intervene in their landscape to enhance the ecosystem, all parts of the ecosystem benefit.

Commensalism is a relationship where one party benefits, but the other is unharmed. An example of humans acting commensally is 'continuous cover forestry' in which a small proportion of the trees in a forest are removed without harming the overall health of the forest.

While the act of continuous cover forestry can be much more sustainable than clearfelling it is also worth noting that we should avoid making explicit claims, and that in reality the answers to most forestry questions are much more complex than we might think. The work of Evolving Forests[3.8] continues to help educate engineers to better understand the complexities of sustainable forestry, and they are continually releasing a new resources.

Parasitism, on the other hand, occurs when one party benefits at the expense of another. A forestry example would be large-scale clearfelling with no plan for regrowth.

Again, while this is often parasitism, there are many natural processes which also clear large areas of forest. The point here is to focus on the act of removing with no consideration for the long-term health of the forest and its inhabitants. But we should avoid applying sweeping generalisations without fully understanding the context of the forest.

For brevity, for the rest of the book we will use the term 'symbiosis' to refer to both mutualism and commensalism, but not to parasitism. In other words we will use it to refer to positive or neutral opportunities, but not ones that do harm.

3.2.2.2 Reflections for the built environment

As we think about how we can work with living systems, we must evaluate the nature of symbiotic relationships. Are they mutual, commensal or parasitic? As we delve deeper into working with renewable systems, here are a few more things to consider:

Rate of renewal: An important factor in renewable systems is the rate of renewal. You can't use up resources quicker than they are reproducing. If you do, you risk destroying the mechanism that is producing the resource. The comparison can be made to spending the interest versus depleting the capital of an investment fund. Start spending the capital and eventually there'll be no fund left.

Living cycles: Living systems, while potentially abundant in their output, do not produce a constant bounty. Growth, harvest, decay and rest all follow a living cycle. Any system that operates at a continuous level of output, probably overlooks these living cycles, and is likely to be using extra energy in the process.

Desertification: When a living system's population falls far below its natural regeneration rate, it enters a state of desertification. In this dormant state, it becomes vulnerable, although it may flourish again if conditions change. Our challenge is to perceive these desertified areas as potentially flourishing systems. Our vision of a different built environment must harness the power of imagination to transform these dormant spaces.

By integrating symbiotic principles into our design, operations and mindsets, we can start to weave our human systems (our supply chains, our energy structures) into the cyclic renewal of the living world. But this idea of symbiosis goes far beyond simply the use of renewable resources. It's a holistic philosophy that recognises the current desertified state of many living and physical systems and recognises their potential for revitalisation.

As we transition from our present state to a more regenerative future, we should be asking: **What can we do to enable the resurgence of thriving, renewing cycles of life that humans can be part of?** Cycles that provide us with the materials and energy we need, not exploitatively, but such that our activities enhance these cycles. And how can we work to reduce our parasitic activities? This approach sets the stage for a construction industry that positively enhances our ecosystems.

3.2.3 Change

The third characteristic of living systems we'll explore is their capacity to change. Living systems exhibit a remarkable ability to self-organise and adapt. By employing simple rules and recurring patterns, they respond to environmental shifts. Those adaptations that best fit the local ecology are successful. This self-organisation in response to local conditions enables a living system to persist and flourish.

We see this principle in action through the land management technique of rewilding. In rewilding, the aim is not to return an environment to a previous state, but to initiate processes that allow the living system to regain its ability to self-organise. The focus of rewilding is on restoring the feedback loops in living systems and letting the ecosystem evolve in a way that best suits its environment (Figure 3.7). Through rewilding, ecosystems can achieve a complexity, resilience and level of flourishing that doesn't happen when humans manage the process. For a detailed description refer to *Feral* by Georges Monbiot[3.9].

Figure 3.7: Since the disaster at Chernobyl 35 years ago, the area has rewilded

3.2.3.1 Reflections for the built environment

This approach to design — creating the conditions in which the system itself can self-organise and adapt — is, of course, very different to how we normally think of design. Our conventional engineering and processes often impose predefined ideas onto a system. We meticulously plan and then execute those plans, often sticking rigidly to them even if the environmental conditions suggest that an adaptation would be beneficial. Rather than allowing designs to emerge organically from the specificities of a situation or environment, we tend to enforce our preconceived notions and established standards.

As Carol Sanford highlights:

> "regenerative design begins from the unique and specific potential of a situation and works to develop the capacity to bring this potential forward, whereas design thinking emphasizes problem-solving as a way of arresting disorder".[3.10]

If our aim is to use how living systems operate as a guide to how we could be designing more regeneratively, then **we are learning that we need to embrace a different mode of thinking; one that aligns with the principles of**

self-organisation and adaptation inherent in the living world. One of the things that enables this self-organisation is the high level of interconnection we have already described. And the specific potential is characterised by what resources are abundant and available in that context. Enabling the conditions in which humans can become more interconnected with living systems will help to create the conditions in which we can co-evolve with these living systems.

Harnessing this capacity for living systems to self-organise **calls for a distinct change in our approach to design; shifting from an imposition mindset, where designs are predetermined, to an emergence mindset, where designs arise naturally from the interaction between our human systems and the living world.** It means thinking in terms of adaptation and response, rather than rigid planning. It's about ensuring our designs not only 'fit' their ecological context but actively contribute to its health and resilience.

3.2.4 Living Systems Blueprint for Regenerative Design

In summary, our aim in this section has been to use living systems as a roadmap for how the construction industry would work if it were truly regenerative. With these principles in mind, imagine a construction industry:

- That is acutely aware of its interconnection with all facets of the human-living system, recognising these links as the foundations of resilience and survival
- That promotes flourishing through integrated, symbiotic cycles of supply
- That seeks to unlock the potential of each part, allowing it to self-organise in response to its environment rather than striving to control it

To visualise this transformation, let's incorporate these principles back into the Systems Bookcase of the future (Figure 3.8). These living systems characteristics can be visualised as threads that run from the bottom of the

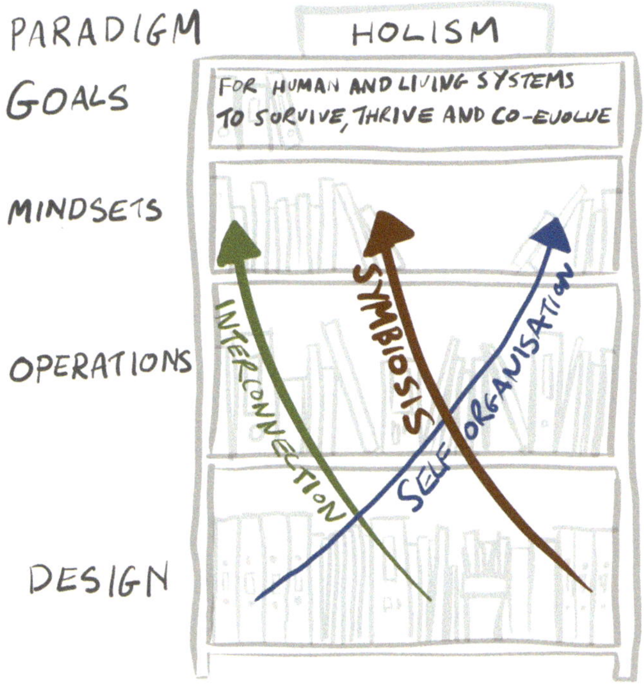

Figure 3.8: Incorporating characteristics of living systems into the bookcase of the future to create the Living Systems Blueprint

bookcase, right to the top — weaving together to achieve the goal of surviving, thriving and co-evolving.

We call this picture our Living Systems Blueprint. It helps us start to visualise how a regenerative built environment industry might look at systems level. It is by no means a complete picture, but we see it as powerful concept sketch that can help guide us.

With the Living Systems Blueprint in mind, we can ask:

- What are we doing to strengthen the interconnection between humans and living systems?
- What steps are we taking to unlock symbiotic resource loops that enable thriving?

- How are we unlocking the self-organising capacity of human and living systems to thrive within ecosystem limits?

With this blueprint of the future in mind, we can start to more effectively plan how we are going to create the transition to that future. How are we going to populate the transition bookcase? What are the blue books we need to write to create the transition? What are the already-existing yellow books that we need more of? What are the red books that we need to remove, and which ones do we want to keep?

In Part 2 (Chapters 4–6) we look in detail at the creation of this transition, using the Living Systems Blueprint as our guide. But before we embark on this journey, the final puzzle piece is to understand the role of the structural engineer in this transition.

3.3 WHAT IS THE ROLE OF THE STRUCTURAL ENGINEER?

In this chapter we are asking: How can we develop regenerative design within a paradigm of continuous growth? The final part of the answer is to consider the role of the structural engineer.

To engage with being regenerative requires us to consider much more than we normally do as structural engineers. This book has already touched upon systems thinking, economics, future thinking and ecology, and there are more topics to follow. Regenerative design requires us to step back and consider a wider picture than we are used to.

Figure 3.9 shows two options for how structural engineers can respond to change. In the upper third a new field emerges adjacent to our world. On the left, where the two fields overlap, the structural engineer shrinks away, enabling this other field to take control of a sector the engineer once understood.

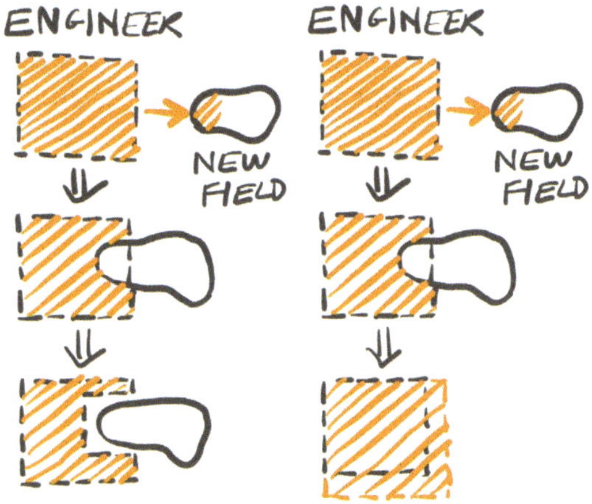

Figure 3.9: The engineer can either withdraw from or embrace new ways of working

On the right, there is the same new field, but this time the structural engineer has expanded their role and influence in order to shape this new dimension of their work. An example of this is the low-carbon agenda, in which structural engineers have quickly adapted to providing advice for clients on carbon.

The IStructE is clear on which way it thinks we should go, as manifested in its guidance, via the UK Joint Board of Moderators[3.11], on how university engineering courses need to be widened in order to cover more topics. This recommendation from the top of the profession encourages us to expand our horizons and assert a leading role in this evolving space.

As we begin our exploration into thinking more regeneratively, it is helpful to start by understanding our current (and future) potential role in the system as structural engineers.

As engineers we never work on projects alone and we don't see this changing as we become regenerative practitioners; if anything the design team will become bigger. Let's consider the different roles we have within this team (Figure 3.10).

Figure 3.10: The different roles we bring to the design team

3.3.1 Enabler

As engineers we first need to recognise that we enable projects. Without a structural engineer a project cannot be safely built, and we are therefore essential to the process. As enablers we take designs and run a number of tests on them; via calculations, models, reviews, discussions, drawing and thinking. These are fundamental skills and as we move into the regenerative design space, we will continue to carry out this role.

As enablers we can start to ask some important questions:

Firstly, what projects do we want to enable? If you are involved in the bidding on projects, do you have a bid/no-bid process? Does this process consider the impacts and opportunity for regenerative design? As a minimum, structural engineers can seek to be regenerative by only working with other regenerative design professionals, although this is a very narrow view of the impact that the engineer can have.

The opposite of enabling is preventing. While individuals can be enablers, it requires industry to come together to prevent projects. Because if one person says they won't work on a project, but others do, then the project hasn't been prevented[3.12].

Following on from this, does your practice have a policy for allowing staff to not work on projects due to ethical grounds? If not, should they? If they do, are you utilising it to make sure that the projects you work on align with your own ethics? Are you enabling projects that you don't want to? Ultimately, every engineer, as an enabler of projects, can decide whether they should be working on a project.

Regularly when we raise this idea, the response from almost every engineer is "but if we don't do this project then maybe someone else will, and we will be able to do it less badly." The problem with this response is that everyone seems to hold the same view, and you can't all be designing less badly than everyone else, and, as we explored in the last chapter, the overall impact of this approach is that we continue to propagate degenerative design.

Another important question to ask ourselves is: Can we enable the outcomes we want to work towards? If a project has the opportunity for reuse, do we have skills to enable this to occur? Likewise, do we have the skills to design with materials beyond steel and concrete? We can only enable projects to happen if we have the prerequisite skills or are willing to learn.

3.3.2 Change agent

As enablers we have agency. If we say something can't be built, people listen, because without us the project can't happen.

Too often on his own projects, James gave up his agency and lost sight of the fact that without his input a project can't happen. But what if, as engineers, we start to insist that projects are not just 'safe enough' but we start to demand that the project, rather than doing harm, does good?

James was recently approached to provide some advice on sustainability on a very large project linked to oil and gas. He declined to work on it, because ultimately the project, while seeking to do less harm, was not in line with his ethics and his desire to see us transition away from fossil fuel use towards a more regenerative future.

Now imagine that we didn't just do this as individuals, but as an industry we were all to do that?

3.3.3 Specifier

As engineers we specify a large amount of the most impactful material on a project. We often relegate the role of specification to the status of dull necessity rather than amazing opportunity. James remembers well the hours spent pouring over hundreds of pages of specification prior to a tender issue, and wondering if anyone would even look at them. But as specifiers of material, we really do have an impact. Specifying material may not require the mathematical acrobatics of design that so many of us enjoy, but hidden in those hundreds of pages are real risks and opportunities that we can work with.

3.3.4 Advocate

As a key member of the design team, we can also advocate outside our primary area of expertise. We may not have the same agency outside the main structural decisions, but we are still valued members of the design team and we can therefore advocate for change and a more regenerative approach across the whole project. Most companies will have relationships with clients, architects and other professionals that span multiple projects and have developed over time. As an advocate these long-term relationships are valuable as they create trust and the opportunity to share new ideas without them being about a specific project.

For regenerative design to be a success — for us to see real reductions in harm and greater flourishing — we need a seismic shift in the way we approach design. This won't happen because a few individuals do it, while the rest of us continue in the existing paradigm, it will only happen if we all (or at least a meaningful proportion of us) do it. So, the role of advocacy should not be underestimated. It will be essential going forward.

3.3.5 Creative

As a key member of the design team, we can also look for creative opportunity. This is especially true at the start of a project, where everything is up for grabs, but remains true throughout the project (although the impact and opportunity reduce). We may not be creative in the same way as other members of the design team, but structural engineers are creative. What opportunities exist and have been overlooked? What questions should be asked, but haven't? Which stakeholders haven't been considered or have been reduced to passive problem-solving roles, rather than encouraged to seek a better outcome for everyone?

3.3.6 Opportunity seeker

As an engineer interested in regenerative design, we should search for opportunities. Opportunities don't just need to come from clients; sometimes the opportunities come from engineers collaborating with others to create ideas that clients then want (refer to the case studies in Chapter 5 for examples of this). Look for the people already demonstrating regenerative practice in their work and ask how you can support them. As we start to explore beyond projects and embrace systems thinking, these opportunities will become more apparent. Opportunities are there, but we won't necessarily find them unless we seek them out.

3.3.7 What about our relationship with the client?

While we are exploring the changing role of the structural engineer, it is worth considering how our client relationship may change. In many large-scale UK-based projects, the client is a developer. But if we want to shift towards a more holistic way of working, we need to have a much broader view of who the client is, that includes the wider human and living system. This clearly challenges the way our work is structured in the current commercial environment, but there is a historical precedence in engineers working for the wider community — i.e., as local authority engineers, and engineers working for the Environment Agency. The idea of wider responsibility of engineers is embedded in codes of professional conduct. Throughout this book you will find examples of engineers pioneering working arrangements that have a very different idea of who the client is.

3.4 CONCLUSION

At the start of this chapter, we asked: How can we develop regenerative design within a paradigm of continuous growth?

We can begin by using the Three Horizons model, to plot our journey from our current design practices to those that promote regeneration and growth. This model allows us to understand the potential transition stages, what we need to let go of in the current system, and what we need to move towards. With this model in mind we can see that a system in crisis can give way to a more viable system by means of creating the right transition stepping stones.

The Three Horizons model helps us see more clearly that one of the biggest barriers to transitioning to a regenerative future is being able to imagine what this future looks like. Fortunately, we have a strong guide to how regenerative systems can operate — living systems themselves.

Living systems, such as woodlands, have been surviving, thriving and evolving for millions of years. Three characteristics in particular can guide us: their interconnection; their use of symbiotic systems; and their capacity to self-organise. Together, these characteristics can help us create what we call the 'Living Systems Blueprint for Regenerative Design'.

Then, at each level in the construction industry system, from design to operations to mindsets, we can ask what we can do to adopt these characteristics in the way that we work so that the construction industry becomes integrated into this living web.

Thirdly, we can leverage all our roles as structural engineers. As enablers, change-agents, specifiers, advocates, creatives, and opportunity-seekers, structural engineers can influence the transition to regenerative design. By asking the right questions, making ethical choices, leveraging their agency, understanding their supply chains, advocating for regenerative design, seeking out opportunities, and creating a different working culture, structural engineers can push the boundaries of the traditional growth paradigm to start to include regenerative design principles.

In summary, we can see that the role of the regenerative structural engineer is to use all their influence at different levels in the construction industry system to gradually expand the regenerative practices we already have in play, and build new transition practices as stepping stones. At the same time, our job is to actively work to reduce the harmful practices currently in use. All this is guided by a deepening understanding of how living systems work.

3.5 CASE STUDIES

The following two case studies are examples of what is called in Three Horizons thinking, 'H3 in the present'. In other words, they are examples that already exist in the present of some of the future we are trying to be build. Don't be deceived: it is not the look and feel of these case studies that we want you to focus on, but the underlying processes that lead to their creation, and the net impact on the living systems they are part of. Examples are necessarily small: the dominant paradigm doesn't (yet) allow practice like this to happen on a bigger scale. But these small-scale projects have an important role to play in challenging our thinking.

CASE STUDY 2

–

Ben Law's Woodland House, UK

In Ben Law's *Woodsman*[3.13] he talks about living in the woods, and starting to coppice sweet chestnut. At first, he coppices as an act of **repair**, and learning. He coppices as the wood has been left unmanaged and uncared for over many years. It could be described as being in a **desertified** state. Through this act he starts to learn about the **rate of renewal**, observing the **living cycles**.

He then builds a simple shelter to enable him to spend more time in the woods learning and observing. As his understanding of the woods and his craft develop, he creates a yurt in which to live. He continues, year-on-year to both observe and work the woodland. Finally, through careful management and developed skill, he is able to produce an A-frame timber house, built with local Roundwood coppiced timber, built by the community that he has developed a relationship with and who are paid in food and camaraderie. The design has emerged from the collective system. The material is suggested by the woods, the form is in response to what is available and is based on traditional forms of construction, and the construction process is discussed and delivered by the community.

This is an example of **mutualism**, where the forest benefits from the actions of Ben Law, and Ben Law receives the harvest which he then turns into a house (Figure 3.11).

Figure 3.11: Ben Law's house was constructed following a long period of observation of the landscape. The design emerged (and continues to evolve) in response to the system

However, this example is also uncomfortable, because for many valid reasons it is not replicable in the current paradigm. It is an example of H3 in the present. That doesn't mean that in the future all houses will be like this one. It is just one example of regenerative practice from which we can learn, while not underestimating the challenges that we would face scaling up this approach.

CASE STUDY 3

–

Tŷ Pren Roundwood construction, UK

If Ben Law's house is an example of H3 in the present. The work of Tŷ Pren is an example of H2 in the present.

Tŷ Pren make Roundwood buildings using locally sourced larch (Figure 3.12). The cooperative was set up in 2014, after one of the founders, Jamie Miller, attended a Ben Law Roundwood timber framing workshop in 2006. Jamie was aware that Ceredigion (an area of West Wales) doesn't contain overstood Sweet Chesnut coppice like West Sussex, and that the local woodland was abundant in unmanaged European Larch.

He explored ways to utilise the larch, which is cut 25 years into the growing cycle as part of the thinning of the woodland. This is an example of **mutualism** as thinning of the woodland enables other trees to grow, and is an essential part of sustainable forestry. However, there is limited economic

Figure 3.12: Roundwood construction and the final completed but unclad frame

incentive for it. By taking the felled larch and turning it into buildings, Tŷ Pren are creating a new supply chain with an abundant waste material which is beneficial to the woods, the woodland owner, and the cooperative who build with the Roundwood timber.

Tŷ Pren work with local forest owners to understand their supply chain. They rent land from one of the owners so that their workshop is setup in the woods where some of their timber comes from. This enables them to **connect** directly with the supply chain and observe, first hand, the impact (both positive and negative) of their work. They are also able to **feedback** to the landowner on the forestry practice, enabling opportunity for more sustainable approaches to develop. In terms of the future planting plan, they are not only looking to plant high quality construction timber species such as Cedar and Douglas Fir, but are planning to plant a Sweet Chestnut coppice that will be managed in a regenerative way to produce the Roundwood needed for the next generation.

Finally, the work of Tŷ Pren was galvanised by changes in Welsh Planning Policy — namely a requirement for low impact homes using the One Planet Living framework[3.14]. As a result, the new homes built in open countryside need to be zero carbon in construction. This created a market for highly sustainable houses. A group of self-employed individuals that were constructing buildings with a natural building ethos, **self-organised** into a cooperative, Tŷ Pren. By being a cooperative, everyone has a voice in key working decisions and everyone is able to hold each other to account (especially those in power) — giving members of the organisation agency.

We noted that this case study is an H2 rather than H3 example. This is because although there are many features of regenerative practice (zero waste, workers' cooperative, storing carbon), ultimately the larch which remains after the thinning will be clearfelled.

The difference between clearfelling and selective felling is not as simple as good/bad. There are a multitude of considerations, not least the type of selective felling. NERR030[3.15] covers a large number of considerations and if you are looking to work directly with a woodland, the advice of a silviculturist should be sought if you want to delve more deeply into this topic.

Therefore while the practice is clearly highly sustainable, unlike the house built by Ben Law which uses the coppiced timber which works within the living systems' cycles, this still leads to a moment of harvesting which will have a significant impact on biodiversity.

REFERENCES

3.1 H3Uni. *Three Horizons.* Available at: www.h3uni.org/tutorial/three-horizons/ [Accessed: December 2023].

3.2 Hopkins R. *From What If to What Next: What if beavers were allowed to reshape our world?* [Podcast] Available at: https://fromwhatiftowhatnext.libsyn.com/66-what-if-beavers-were-allowed-to-reshape-our-world [Accessed: December 2023].

3.3 IPBES. *Global Assessment Report on Biodiversity and Ecosystem Services.* Available at: www.ipbes.net/global-assessment [Accessed: December 2023].

3.4 Meadows DH. *Thinking in Systems.* London: Chelsea Green Publishing, 2015.

3.5 Wohlleben P. *The hidden life of trees.* Glasgow: William Collins (HarperCollins Publishers), 2017.

3.6 Rebanks J. *English Pastoral: An Inheritance.* London: Allen Lane (Penguin Random House), 2020.

3.7 Smith P. *Trees: From Root to Leaf.* London: Thames & Hudson, 2022.

3.8 Evolving Forests. *Looking beyond the usual solutions to strengthen our connection with trees.* Available at: https://evolvingforests.com/ [Accessed: December 2023].

3.9 Monbiot G. *Feral: Rewilding the Land, Sea and Human Life.* London: Penguin, 2014.

3.10 Sanford C. *The Regenerative Life.* Boston MA: Nicholas Brealey Publishing, 2020.

3.11 Joint Board of Moderators. *Guidelines for Developing Degree Programmes (AHEP3).* Available at: www.jbm.org. [Accessed: December 2023].

3.12 Turner S and Norman J. Own your agency: what part will you play in the climate revolution? *The Structural Engineer.* 2021:99(3). Available at: https://doi.org/10.56330/JQDH9427 [Accessed: December 2023].

3.13 Law B. *Woodsman.* London: HarperCollins Publishers, 2014.

3.14 Bioregional. *One Planet Living®.* Available at: www.bioregional.com/one-planet-living [Accessed: December 2023].

3.15 Natural England. *Environmental impacts of land management (NERR030).* Available at: https://publications.naturalengland.org.uk/publication/30026 [Accessed: December 2023].

PART 1 SUMMARY

In Part 1 we have used a systems lens to explore how the construction industry operates, and how, despite our ambition to be sustainable, our industry continues to cause net harm to the living systems that we depend upon. We have established regenerative design as a new goal for our industry, set within a paradigm of holism, in which the aim of our work is for human and living systems to survive, thrive and co-evolve.

While this holistic way of thinking is strongly at odds with the traditional construction industry mindset, we have seen, using the Three Horizons model, how one system of working can give way to another. The crucial step is an effective transition from the world that we see, to the world we want to build. Key to unlocking that transition is having a clear understanding of what it is we want to create. To help map the way, we established the Living Systems Blueprint to guide us in thinking about how we can create a transition away from current practice towards a regenerative built environment sector.

Finally, we saw that the structural engineer fulfils many roles in design teams, in industry and in communities. All of these can be leveraged to create the change we seek.

In Part 2, we build on this groundwork to look in detail at how we can create that transition at the level of design, operations and mindsets.

Part 2
Creating the transition

In Part 2 we explore in depth the idea that the role of the regenerative structural engineer is to create a transition to a regenerative construction industry. Each chapter explores how we can facilitate this transition through a different aspect of our work.

In Chapter 4 we investigate how we can use design to help us transition to a more regenerative industry.

In Chapter 5 we delve into how structural engineers can facilitate systemic change at the operations level.

In Chapter 6 we look at how we can create the mindset for a transition to a more regenerative industry.

CHAPTER 4

Creating transition through design

In this chapter we are going to focus on building a transition framework. In doing so, we are looking at the bottom shelf of the transition bookcase (Figure 4.1). We'll consider four key stages in design: observation, briefing, developing ideas, and making and testing. Across these stages we'll be asking what we can do to build interconnection between human and living systems, where can we unlock healthy symbiotic relationships in these systems, and how do we release the self-organising potential of the places we are building in?

In short, **how can we use design to help us transition to a more regenerative construction industry?**

Figure 4.1: Creating a transition to regenerative practice through design

Primarily, we frame this chapter at project level, as this is where we believe most readers will feel the most agency. However, as will quickly become evident, regenerative practice involves a wider perspective beyond individual projects — it includes the broader construction industry and its interaction with human and living systems.

In each section we consider first what would ideal regenerative practice look like, and then what might be the practicable steps we can take now to challenge the existing system and shift practices towards more regenerative ways of working.

4.1 OBSERVATION

The Living Systems Blueprint for Regenerative Design shows that we are dealing with a much more complex and interconnected world than we might usually consider as structural engineers. Because of the unpredictability of complex systems, we need to change our approach, beginning with an extended period of observation to try to understand how the system operates before intervening. But what does this mean in practice?

Of course, any design process involves a certain amount of research. What does the client want? What do the stakeholders care about? What are the site conditions? Observation is just another kind of research, but here it goes much deeper, and has multiple aims:

- It aids in recognising and understanding the interconnection between the project, the site, stakeholders, and the broader living world
- It assists in identifying potential opportunities for symbiosis within these connections
- It provides a deeper understanding of the specific potential of the site and what the best-fit evolution for that place might be

Through observation, let's look in turn at the ecosystem, the materials, the community, and construction processes.

4.1.1 Ecosystem

We begin our observation by trying to develop a deep understanding of the local ecosystem. Here are some starting points:

- **Biodiversity:** Investigate the different forms of life in the area. Tune in to the various species — trees, birds, insects, plants, fungi etc.
- **Life-enabling factors:** Analyse what is currently enabling life to flourish in the environment. This could include specific habitat conditions, availability of food sources, or certain climatic factors
- **Gap analysis:** Identify what seems to be missing or scarce in the ecosystem that might otherwise support a richer biodiversity. Visualise what the place would look like if life were flourishing. Try to understand what constraints or obstacles might be preventing life from flourishing
- **Physical characteristics:** Examine the topography of the area. Where does the water naturally flow? Which areas receive more or less sunlight? These factors significantly influence both the ecosystem and potential construction
- **Symbiotic opportunities:** Look for opportunities to work symbiotically with the ecosystem. This might be through mutual symbiosis (both parties benefit) (Case study 4) or commensal symbiosis (one party benefits, the other is unaffected)
- **Cycles:** Identify the living and physical cycles present in the ecosystem. These include seasonal cycles, day-night cycles, tidal cycles if near a body of water, or even life cycles of various species. How do the seasons affect the environment, its habitats, and its species? Are there patterns of migration, production, or other significant

changes? Recognise how these cycles affect the ecosystem and how they could be worked with (Case study 5)

All these are a starting point for immersing ourselves in the ecosystem to understand how life flourishes there. Observation of this sort **shifts us from viewing the living world as a natural backdrop to seeing it as a dynamic, evolving system of which our design is part**.

This observation process should take us much deeper than a traditional desk study. The emerging science of 'embedded cognition' shows that we understand the world by moving through it and experiencing it[4.1]. So, engage your senses! And find out what you can learn from the local community and if possible, from Indigenous voices from that place, who may be more connected to this ecosystem.

This act of observation should be one of active immersion and continuous learning. This deep immersion in the local ecosystem will help us to see our projects as not placed 'on' the land but as becoming an integral part of the living systems that already exist there.

Of course, we need to recognise that deep immersion is not always possible for everyone, as noted in a recent article in *It's Freezing in LA*[4.2]. And even when we do spend time looking, we still miss things, as noted in *Addis Ababa: A Manifesto on African Progress*[4.3]. We should also consider adding another set of eyes to the observation process, ask them to record what they see; be that someone from the local community, or someone with a different focus.

CASE STUDY 4

Home Grown Cabin, UK

The Home Grown Cabin by London Metropolitan University and Grown in Britain, with engineering support from Mule Studio (Figure 4.2) is an example of observing the eco-system and finding symbiotic opportunities. In this case, the cabin utilises coppiced timber, an example of mutualism, as the coppicing of sweet chestnut enables the local ecology to thrive as it gains access to sunlight. It also includes the use of ash dieback timber, which is typically burnt. By utilising it in construction the carbon is stored long-term, in the building fabric.

Figure 4.2: Home Grown Cabin

CASE STUDY 5

La Vallette Bathing Pools, Guernsey, UK

La Vallette Bathing Pools in Guernsey (Figure 4.3) are a set of Victorian pools which fill with seawater at hightide and capture this water for bathers, providing a safe and comfortable environment. They are an example of integrating physical cycles into design. The pools were recently restored and a new changing facility and café were constructed. The project was funded by the local community and the pools are free to use.

Figure 4.3: The bathing pools

4.1.2 Materials

As we now shift the lens of observation to think about materials, we will invoke the concept of the 'second site' that includes all the places in the world affected by the project's supply chain and waste streams. The consequences of our actions on this second site may far exceed the impacts we can see on the construction site. The challenge is that we don't normally see these second sites; the places where we have impact. With this idea in mind, we can consider:

This idea is based on the concept of the 'second body' (pages 78–79) from Daisy Hildyard's book of the same name.

- **Material origins:** Where do the materials on the project come from? What is the wider impact of their extraction on the ecosystem and community of origin? Are these systems going to be enhanced or depleted through the use of the material on your project?
- **Material renewal:** Is the process of material production renewable or extractive? If renewable, is the rate of extraction sustainable? If extractive, is there an abundance of supply?
- **Existing materials:** Could existing buildings be used as a supply of materials for new ones? What surplus materials exist in the locality? What could be repurposed? What circular material supply chains already exist? Are there local sources of materials that could be rejuvenated or grown? Could the existing materials' life be extended through investigation? (Case studies 6 and 7)

The second site

The term 'second body' coined by Daisy Hildyard[4.4], encourages us to think about not just the physical space that we individually inhabit, but the wider space that we touch through our consumption and waste. This second body can contribute positively or negatively to wider society and the environment. While it is much harder to take responsibility for this second body, its impact may far exceed that of our own.

We can apply the same logic to construction projects. For every construction site there is a 'second site', the wider locus of the impact of the construction supply chain and our waste streams. This second site could extend globally. In fact, we already do this with CO_2 — the emissions that we consider in the embodied carbon calculation, mostly occur offsite.

Of course, through responsible procurement practices we can mitigate the negative impact of our supply chain. But do we really get a rich understanding of the impacts of our decisions? For example, does sustainably sourced timber lead to greater biodiversity or simply promote more woodland monoculture?

Understanding this 'second site' requires observing the entire supply network, connecting ourselves with the origin of the materials. If this sounds preposterous, then it's about adjusting our perception of what is reasonable. Just as carbon counting has become commonplace, we should expect to develop as deep an understanding of the places our materials come from as the places in which we build.

Recently we visited Avonmouth to see the docks through which tonnes of the building materials we use every day in our industry are shipped. We were, understandably, unable to enter. But even peering in from the perimeter we could feel it on our eyes. We could taste it. For years we have specified concrete and never felt it on our eyes, never once tasted it.

Figure 4.4: A view into Avonmouth Docks

The reality is that we have disconnected ourselves from industry as much, if not more so, than from the living world. We may go outside for walks and runs but we rarely visit the industry on which our life depends. Our first challenge as engineers is to observe it. As specifiers of materials how often do we add to the specification that we would like to visit not just the fabrication shop but the manufacturer and the source of the materials? This could have many positive impacts. Firstly, we can see where our material is coming from. Secondly, the closer it is the less distance we will need to travel. Thirdly, the idea that someone may visit the manufacturer might challenge the subcontractor to consider the supplier more carefully (of course we hope they already do, but it will add an extra incentive).

Once we start to think about the providence of our material, once we observe the impact and start to understand it, we can start to consider how to do things differently.

- **Material transport:** How does the material get to your site? What is the impact of this journey on the places in between? Not just the carbon footprint but the wider impact of all the logistics. This could include social disruption and the arrival of invasive species
- **Waste streams:** What waste streams from other processes could become inputs to your project? And what happens to the waste from your project?
- **Connection to suppliers:** Building connections with people supplying materials is vital. These relationships not only ensure an understanding of the materials' origins and impacts, but also open opportunities for mutual learning and discovery

These are a starting point for observing the processes and impacts related to material supply. Again, the deeper level of understanding comes when we make physical contact with as many of these processes as possible. Engagement at this level will help us to see our design as acting not just on a site but on all the places we touch — from material origin to finished building.

CASE STUDY 6

Elizabeth Line OLE Gantry, UK

In 2022, Elizabeth Line OLE Gantry Rigorous Assessments won the IStructE's Structural Award for Minimal Structural Intervention. The project is an excellent example of how deep **observation** of the existing structure, followed by rigorous analysis that looks to understand design beyond the limitations of our codes, can lead to substantial **reuse** of structures (Figure 4.5). However, in conversation with the engineer we discovered that the approach taken has now been more widely adopted by the client, Network Rail, and has been captured in new design guidance which is available to other engineering practices. This shared information could lead to a system change, where gantries rather than being systematically replaced are now assessed, repaired and reused.

Figure 4.5: Elizabeth Line OLE Gantry Rigorous Assessments

CASE STUDY 7

–

Reuse of blockwork, Tanzania

In Tanzania, most buildings are constructed with hollow cement blocks measuring approx. 460 × 230 × 125mm. The roofs consist of wooden or iron beams covered by corrugated iron sheets. Additionally, the blocks are systematically arranged, resulting in a wall thickness of approx. 140mm (including the plaster width).

The buildings in Figure 4.6 are located in Dar es Salaam. In this geographical area buildings are often damaged by flooding. Rather than demolish buildings after the damage, the buildings are dismantled and the hollow concrete blocks are checked and those that are not damaged are stacked up to be reused for new constructions (Fig. 4.6b).

a

b

c

Figure 4.6: Reuse of hollow blocks in Tanzania

4.1.3 Community

Our third observational lens is through the communities that we work with. As with materials, the communities that our design work impacts extend far beyond the locality of the construction site itself. Our goal is for human and living systems to survive, thrive and co-evolve. This means we need to spend some time understanding these human systems. Here are some things to consider:

- **Connectivity:** Identify the people that the project affects and find ways to hear their voices. Try to understand the link between people and the places they inhabit. This extends to the places touched by our engineering supply chains. What is it that binds the community together? What connections give the community its resilience? Also try to observe how this community connects to their wider ecosystems. On what local resources do they depend?
- **Inequality:** Essential to creating thriving human and living systems is seeking to understand the drivers of inequality in a community. This could be inequality of health, income, education or opportunities — longstanding or recent. All these observations will help seed ideas for how a design can contribute to healing of human and living systems
- **Abundance:** What is the community rich in? Not just in monetary terms but in culture, heritage, knowledge and skills. What symbiotic relationships support this wealth and how could you work positively with these?
- **Local knowledge:** The local community may possess knowledge of how to work with local materials. There may be deep knowledge of the local landscape, the soil, what grows there, how the water flows. Local culture may be steeped in historic trading of materials,

reflecting ways of working that may be specific to that place that you can tap into

- **Future generations:** Part of the work of observation is to look at how the existing built environment infrastructure might begin affecting the prospects of future generations, and how might intervening in that community change those prospects. This concept is at the heart of the Welsh Government's Well-being of Future Generations Act[4.5]

These considerations encourage us to take a deep dive into the potential of the local community and, as we trace the web of our supply chains, the potential of the communities in locations around the world. Again, **this approach encourages us not to think of the community as bystanders to our projects but a dynamic system of which our design is part.** Deep immersion in the community will help us develop designs that rather than being irrelevant to the community and its heritage, are woven into the social and cultural systems already there (Case studies 8 and 9).

CASE STUDY 8

–

Westonbirt Community Shelter, UK

One of the most beautiful examples of local community engagement that we have seen comes from Westonbirt Arboretum. In the design and build of this community shelter (Figure 4.7), the design was created so that not only were the materials locally sourced, so was some of the labour. The project, which provides shelter for a number of community groups, was constructed by some of the final users, including a number of disabled users.

In addition, the structure, which is a timber grid-shell with a timber shingle roof, sources much of the timber directly from the arboretum where the shelter is constructed.

Figure 4.7: Westonbirt Community Shelter

CASE STUDY 9

–

Gabiro Agribusiness Hub, Rwanda

In Rwanda, East African Piling (EAP) were asked to install piles for a major irrigation project. Rather than import their workforce (EAP are based in Uganda) they worked with local staff, with over 90% of the team being Rwandan, half of whom came from the local village. Many of the staff had never heard of a piling rig, let alone worked with one, so EAP provided training and at the end of the project (Figure 4.8) they organised a barbecue and awarded each member of the team a certificate of employment and participation on the project. The training of local staff means that the impact of the project goes beyond the water supply as it creates further employment opportunities.

Figure 4.8: The piling team for Gabiro Agribusiness Hub celebrating

4.1.4 Local construction

Our fourth observational lens is through the construction of local buildings and infrastructure.

- **Vernacular design:** How have buildings in the geographical area been designed and built in the past? What does that suggest about the local culture, economy, materials and climate? Historical construction choices may be much more closely tuned into the necessities of a place than more modern buildings built in the time of cheap energy (Case study 10)
- **The state of buildings:** Evaluate the state of existing buildings. How are they built? How are they responding to shifting climatic conditions? What is their expected lifespan?
- **Traditional materials:** What building materials are traditionally employed in this area? How do they weather and decay over time? Where do they come from? What knowledge is needed to work with these materials? Is there a modern equivalent that leverages these traditional methods?
- **Construction techniques:** Investigate the construction methods used in the area. Is there a technique specific to this location? There may be a local approach that is particularly resilient or sustainable. Identify any construction methods that may have been lost over time. Why have they been lost?
- **Environment:** How have buildings been designed to manage the local environment? What strategies are used for cooling and warming, managing sunlight, responding to the wind, humidity, air quality?

CASE STUDY 10

–

Rebuilding after 2005 earthquake, Pakistan

Yasmeen Lari, winner of RIBA's 2023 Royal Gold Medal, talks about the moment her approach to building design went through a paradigm shift. After the 2005 earthquake in Pakistan, she felt a need to help. What she discovered was an international response that took no consideration of the locally available materials or the vernacular of the area, and instead provided prefabricated solutions. And so, with locals, she used the readily available mud and stone from locally collapsed buildings, and started the work of rebuilding[4.6] (Figure 4.9). However, as was evidenced by the destruction from the earthquake, updates to the design were required, so she worked with engineers to create a new vernacular, still using local materials and forms, but now with the addition of ring beams to provide structural stability in future earthquakes. Lari has since gone on to design a number of beautiful buildings using local materials and worked with communities to provide safer cooking stoves which empower women in the local community[4.7].

Figure 4.9: To create long-lasting buildings the local vernacular is adapted to create seismic-resilient homes

4.1.5 Conclusion: observation

Building deeper observational practice is an important transition step to regenerative practice through design. From deep observation of the ecosystem, our supply chains, and our community and construction processes, we extend our understanding of the impacts of our work far beyond the traditional boundaries of a construction site. Two important factors emerge from this:

Firstly, transitioning to more regenerative practice requires us to develop deeper relationships with both the place and the community. **The richer our understanding, the greater our capacity to create designs that are embedded in and are an extension of the existing social and living systems.**

Secondly, we must recognise the complexity that comes with extended supply chains. The further these reach, the more challenging it is to fully comprehend the ramifications of our actions, which in turn makes regenerative design more challenging. **Transitioning to more regenerative design practice involves bringing more of the supply chain into view so that we can more fully assess the impact — both negative and eventually positive — of our design decisions.**

As we go on to explore the next stage in the design process (Section 4.2) we'll be thinking about how we can respond to the interconnection and complexity we have found in the observation phase.

4.2 BRIEFING

Following the observation phase, we now need to develop a design brief. The brief can be anything from a rough set of intentions for a piece of collaborative work to a formal document defining client specifications for a development.

In the design process, we see the brief-writing phase as the most powerful opportunity we have to create a transition to more regenerative practice in construction. This is because it is much easier to influence the impacts of a project at early-stage. At the level of design, writing the brief is our greatest opportunity to reinforce existing regenerative practice in the construction industry, to establish transition approaches that are a stepping stone to regenerative practice, and to reduce practices that we want to phase out.

To guide in this process, we have defined four principles (Sections 4.2.1–4.2.4) for writing and interpreting a design brief that can help us transition towards more regenerative practice in construction.

4.2.1 Aim higher in the system

Our first principle, aiming higher in the system, encourages us to extend our vision beyond the boundaries of isolated projects, striving for systemic change that enhances the capacity of human and living systems to survive, thrive, and co-evolve. This objective can't be achieved by examining individual projects in a vacuum. Instead, we must see each project as a catalyst for wider systemic transformation (Figure 4.10).

When we start to see design in this light, we are starting to think more holistically. **The key step is to shift from thinking about designing standalone buildings towards seeing design of buildings and infrastructure as a catalyst for wider systemic change.**

Through our projects, we hold the power to shape the broader systems at play. We can influence, for example, supply chain organisation, interpretation of local planning laws, and development of governing codes of practice. These actions represent operational-level interventions (acting at the operations level of the Systems Bookcase),

Figure 4.10: Aiming higher in the system

into which we will delve deeper in Chapter 5. We can drive research and innovation, shape organisational culture, and even transform shared values. These actions represent mindset-level interventions (the next level up in the Systems Bookcase) influencing the deep-seated beliefs that underpin our industry. We will explore these ideas further in Chapter 6.

This transition in thinking moves us from asking: How can I design this structure more regeneratively? to posing the question: How can the act of designing this structure create a transition to more regenerative systems in construction? (Case studies 6 and 11).

The next three principles will provide guidance on how these overarching ideas might translate into tangible design briefs.

CASE STUDY 11

–

Holbein Gardens, UK

Heyne Tillet Steel (HTS) were the engineers on Holbien Gardens, a project that, in their own words:

> "is the refurbishment and extension of an existing 1980s office building with sustainability at its heart. 93% of the existing building fabric has been retained and the new two storey extension is a hybrid construction of engineered timber and reused steel. The proposed steel and CLT extensions will incorporate 11 tonnes of existing steelwork found onsite as well as steel sections taken from demolition sites across the Grosvenor portfolio (Figure 4.11).

Figure 4.11: Installed steels showing holes from previous uses

> Steel sections reused in this way have around 1/8 the embodied carbon of traditionally-sourced steel. The project successfully demonstrates how circular economy principles can be applied to projects in practice, and if scaled up could make a significant improvement on embodied carbon for steel frame construction."[4.8]

However, one of the key aspects of the undertaking was not just to deliver a project, but to share the learning that occurred, and to enable other practices to carry out similar designs. This is a clear example of setting a brief that aims higher in the system.

HTS have shared the project in many forums, including an article in *The Structural Engineer*[4.9], a number of blog posts about the process[4.10] and presentations on the project. This information enables other engineers not just to be challenged by the work, but shares the learning to enable them to also carry out similar work.

4.2.2 Aim to re-establish interconnection

We move on to our second principle for writing and interpreting a brief in a way that helps us transition to regenerative practice; aim to re-establish interconnection. This principle recognises our work doesn't exist in isolation. Instead, we aim to re-establish interconnection between ourselves as designers, the community, the ecosystem and the place itself. We can find opportunities to create interconnection through the many roles we fulfil: as enablers, change-agents, specifiers, advocates, creatives and opportunity-seekers. This principle provides a different perspective on design, prompting us to redefine traditional parameters and to create a more interconnected approach:

- **Change the line around the site:** As we saw in Section 4.1, the real impact of conventional design projects extends far beyond the site boundary, encompassing areas influenced by the supply chain and other associated activities. To transition to more regenerative practice, we need to write and interpret briefs in a way that extends the 'site' to include more of the places where we have impact, to include the 'second site'
- **Interconnection with ecology:** Four simple measures of ecosystem health are water quality, air quality, soil health and biodiversity count. Creating a brief that aims to increase all four of these measures compels the design team to remain connected to these aspects of ecosystem health and helps us transition towards a design in which we fully account for our impact on the ecology of the site
- **Interconnection with community:** How can the project foster a deeper interconnection between the community and the buildings they inhabit? Between the people who build a home and the people who live there? Can this interconnection be a fundamental aim of the

project rather than a supplemental feature? Consider ways in which the project could contribute to bridging social injustice, fostering unity rather than reinforcing divisions (Case study 12)

- **Link community and ecosystem:** Consider ways the project can establish a greater interconnection between the community and the ecosystem it is part of. For example, can the design create the situation in which the project not only builds a relationship of stewardship between the community and the local ecosystem, but one in which the community directly benefits?
- **Long-term relationships:** Aim to create a brief that encourages the formation of enduring relationships with the site and its community. As we've previously discussed, human and living systems are complex. Fully understanding the impact of our activities requires engagement over extended periods. This could, for example, translate into contracts for long-term maintenance or monitoring, or community partnerships that extend beyond the project's immediate lifespan. In this context, the role of a cathedral engineer, who cares for a structure over decades, is a good example (Case study 13)

In summary, writing a design brief is an opportunity to transition to more regenerative practice by developing interconnection between the design team and the ecosystem we are impacting, within the ecosystem and its community. In this way, each project is a potential catalyst for growing feedback within the system, and creating the conditions for beneficial symbiotic relationships to emerge (Section 4.2.3).

CASE STUDY 12

–

Steel Warriors Outdoor Gym, UK

Heyne Tillet Steel were the engineers on a project called Steel Warriors[4.11]. The project takes knives that have been confiscated or handed in and melts them down. The steel is then converted into outdoor gyms (Figure 4.12), which are located at strategic points that intersect different gang territories to enable communities to come together and restore relationship. This is an example of how an engineer **enables** a project to occur, through testing and design, and also of how, through **connection with community** you can bring about community healing.

Figure 4.12: Example of a Steel Warriors Outdoor Gym

CASE STUDY 13

—

Wells Cathedral, UK

In the UK, every cathedral is required to have an architect. Their job is to understand everything about the cathedral. Some cathedrals, such as Wells Cathedral (Figure 4.13), also employ an engineer. They don't employ a practice, they employ a person and this person, once appointed, is expected to remain in post until they retire. One of James's former colleagues took exactly this role. The incumbent engineer passed on their knowledge and wisdom so that when the appointee took over, they already had a reasonable understanding of the building and its structural form, so that future opportunities could be realised and future threats could be resolved. Cathedrals are often centuries old (work started on Wells in 1175) and engineers and architects are required to play a part of the much bigger continuous project. They can help direct the long-term investment strategies (of both time and money) and ensure it considers the needs of the future. There are occasions where work is required due to an increased risk to the structure, and others where work is required due to an opportunity. But all of this is born out of knowing the project and having a **long-term relationship** with it.

Figure 4.13: Wells Cathedral

4.2.3 Aim to build symbiosis

The third principle, building symbiosis, shifts our focus towards resource supply. A transition step towards a regenerative construction industry is to move away from purely extractive practices towards creating mutually beneficial symbiotic loops in our materials and energy use. We can help to enable this transition in design by writing a brief that seeks to unlock the potential of symbiotic systems. Let's explore this further:

- **From consumption to production:** Flip the script of the design brief. Instead of solely focusing on the consumption of resources, aim to cultivate and expand a material resource. For instance, the development of Modcell®, a standardised straw-bale panel for construction, represents this concept well. The objective was not just building with straw-bales but establishing a process to transform locally grown crops into a scalable industry (Case study 22, Chapter 5)
- **Create the conditions for symbiosis:** Another approach is to create the conditions within which symbiosis can emerge. For example, designing buildings for deconstruction and disassembly and creating a platform for construction material exchange (Case study 14) or creating guidance that enables other people to use a renewable resource (Case studies 6 and 11)
- **Restoring abundance:** Some living systems are significantly depleted, existing in a desertified state. Before engaging in a symbiotic relationship with them, they must be restored to a state of health. We can see this restoration as a regenerative design brief in itself, as it is helping to create the conditions within which humans can work with living systems to meet our needs
- **Work with regional variation:** Traditional construction methods made use of locally abundant materials. Modern structural engineering, in contrast, works

predominantly with standardised components. Transitioning towards more regenerative practices involves making the most of regional abundance in materials and skills, similar to traditional practices

- **Unlock community capacity:** Renewable resources encompass more than just materials. Aspects like trust, community spirit, and a sense of belonging and purpose represent renewable human capacities. These capacities are fundamental to human thriving[4.12]. Therefore, another potential aim for a brief could be to foster an increase in these human capacities as a result of the project (Case studies 15 and 16)

Of course, these considerations don't exist in isolation. Simple interventions can yield multiple, stacked benefits. For instance, planting street trees enhances living habitats, moderates street temperatures, shades buildings, improves wellbeing, and one day provides construction materials (Case study 19).

Overall, the creation of symbiotic resource cycles builds on the interconnections we described in Section 4.2.2 to create a construction industry that through its connections with human and living systems, can thrive within its ecosystem boundaries.

The final aspect to consider, is how to create a brief that helps us to transition towards co-evolution of human and living systems.

CASE STUDY 14

—

Nationale Bruggen Bank, Belgium

Nationale Bruggen Bank (National Bridge Bank) is a Dutch organisation that reuses bridges and bridge components (Figure 4.14). The bank includes a catalogue of available resources, creating a platform for construction material exchange[4.13]. It also includes a guide that enables others to understand how to reuse bridges. This is an excellent example of **creating the conditions for symbiosis** and can also be seen, through the sharing of information, as a way of **aiming higher in the system.**

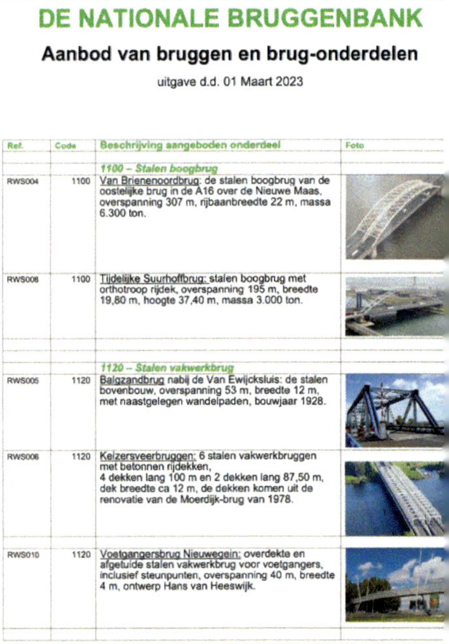
DE NATIONALE BRUGGENBANK

Aanbod van bruggen en brug-onderdelen

uitgave d.d. 01 Maart 2023

Ref.	Code	Beschrijving aangeboden onderdeel	Foto
		1100 – Stalen boogbrug	
RWS004	1100	Van Brienenoordbrug: de stalen boogbrug van de oostelijke brug in de A16 over de Nieuwe Maas, overspanning 307 m, rijbaanbreedte 22 m, massa 6.300 ton.	
RWS008	1100	Tijdelijke Suurhoffbrug: stalen boogbrug met orthotroop rijdek, overspanning 195 m, breedte 19,80 m, hoogte 37,40 m, massa 3.000 ton.	
		1120 – Stalen vakwerkbrug	
RWS005	1120	Balgzandbrug nabij de Van Ewijcksluis: de stalen bovenbouw, overspanning 53 m, breedte 12 m, met naastgelegen wandelpaden, bouwjaar 1928.	
RWS006	1120	Keizersveerbruggen: 6 stalen vakwerkbruggen met betonnen rijdekken, 4 dekken lang 100 m en 2 dekken lang 87,50 m, dek breedte ca 12 m, de dekken komen uit de renovatie van de Moerdijk-brug van 1978.	
RWS010	1120	Voetgangersbrug Nieuwegein: overdekte en afgetuide stalen vakwerkbrug voor voetgangers, inclusief steunpunten, overspanning 40 m, breedte 4 m, ontwerp Hans van Heeswijk.	

Figure 4.14: The Nationale Bruggen Bank catalogue of bridges for reuse

CASE STUDY 15

—

Bridges to Prosperity, Rwanda

Bridges to Prosperity (B2P) builds bridges that are constructed predominantly by the local community who will benefit from the new bridge. This creates a motivated workforce, and it also means that every element of the bridge is understood by someone in the local community. The international volunteer team and the B2P representatives add knowledge and direction where possible to educate the workforce about the construction and maintenance of their bridge. At the end of the construction phase, several key individuals from the local build team are identified and formally given maintenance and repair roles and responsibilities. The local community knows who these individuals are, and in accordance with the local Rwandan culture, they are respected and relied upon (Figure 4.15). The result of this process is to embed the health and maintenance of the bridge into the local community. If they don't look after their bridge then they won't benefit from it. The charity also keeps an open communication with the community and if a significant repair is required, they will aim to help.

Figure 4.15: Bridge in Rwanda built by the local community

CASE STUDY 16

–

WeCanMake, UK

WeCanMake is a community-led response to the housing crisis led by the Knowle West Media Centre[4.14]. In their own words:

> "WeCanMake has a people and place-led approach to housing innovation, embedded in and working with the community of Knowle West, Bristol (an area of high deprivation). Our mission is to grow the spaces, tools and capacities to collectively imagine and actively anticipate better community-led ways to do housing.
>
> So far, WeCanMake has delivered two community-led, low-carbon, locally-made, Living Rent homes in Knowle West, with the land and homes held in trust for community benefit in perpetuity. Our two homes are a practical demonstration of a new replicable model for community-led, opt-in densification, that unlocks an additional supply of land enabling those in housing need to create affordable homes precisely where they are needed most."

Price and Myers has helped to enable this community design-and-build project through a novel oriented board strand (OSB) construction method called BlokBuild®, which they have been developing as part of an ongoing research and development project (Figure 4.16). Recent developments have included replacing steel nail plates with OSB-based fixings that have lowered the system's carbon intensity and simplified assembly onsite.

As well as this being an example of structural engineers helping **to unlock community capacity**, they have also shared their approach with others, enabling other communities around the UK to use a similar template designed to suit their own locality[4.15].

Figure 4.16: Constructing an OSB house

4.2.4 Aim to enhance capacity to self-organise

Our final principle is to write or interpret a brief in a way that helps us transition towards a more self-organising approach to how the construction industry interacts with the living world, and, more widely, how human and living systems interact with each other.

It is about using our abilities as structural engineers to facilitate all elements within a particular site to collectively create the resources and habitats — including buildings and infrastructure — that we all need. This idea challenges the very way we think about design. Let's recall Carol Sanford's words in Chapter 3:

> "regenerative design begins from the unique and specific potential of a situation and works to develop the capacity to bring this potential forward, whereas design thinking emphasizes problem-solving as a way of arresting disorder"

This observation reminds us that regenerative design commences from the distinct potential of a situation and strives to realise this potential — that as structural engineers we should be potential seekers, not just problem solvers. To help us transition towards this self-organisational way of thinking, we suggest asking the following three questions when writing or responding to a brief:

These questions are inspired by a workshop given at the Regenerative Design Lab by Cíaran Malik on looking for patterns of place.

- What are the repeating patterns in:
  - the place I am working in?
  - my supply chain?
  - the tasks I'm carrying out?

- What is enabling these patterns (and what is preventing them)?
- What is the most appropriate evolution?

Let's apply these questions to four different contexts: society, living things, buildings and transport:

- **Society:** What do people usually do there? Is this a place of gathering? Is this a place of trade, of meeting, of living, of manufacture? What is it about this place that has encouraged this behaviour? What have people stopped doing? What has dwindled? What has caused the decline? How do we best organise for society on this site?
- **Living things:** What grows here and why? What things call this place home? What are the specific habitats that have enabled this specific collection of living things to live here? What used to be present but has now dwindled or is gone? What were the factors forcing the decline? How do we best organise this site to support living things?
- **Buildings:** How are things built here now? How has that changed over decades or centuries? How are the buildings laid out? Has that arrangement remained static or changed? What building materials were used and why? What historic methods of building have disappeared and what remain? What is the best evolution of buildings on this site?
- **Transport:** What travels through this place? People? Living things? Minerals? What drives travel, and what steers things here? What did these travellers bring and leave behind? Who or what no-longer comes here? What caused this diversion? How do we best organise this place of transport?

Ultimately, writing a brief that helps us transition towards more self-organised ways of working requires us to be empathetic to and curious about the site and its potential. It requires a transition in our thinking towards being collaborators with human and living systems (Case study 17).

CASE STUDY 17

–

Atelier Luma, France

Atelier Luma is a fascinating example of what happens when a brief is set that enables the designers to work with the potential of site rather than imposing a design on a place. On the surface, Atelier Luma is a reuse project — taking an existing train depot and converting it into a design and research laboratory, 'including workshops for timber, metal, ceramic and textiles alongside dedicated algae and mycelium laboratories, meeting rooms, desk and production space, library and a resource centre'. But what interests us is the way the project began. 'The process began with mapping the region's resources, industries and waste products, identifying streams of both materials and local know-how'[4.16].

We see this process as the designers seeking to work with local patterns and the potential of place.

The building includes a number of innovations looking to use locally sourced bio-waste materials. It includes rice straw-bales for insulation and sunflower seed husks for acoustic insulation. Rammed earth walls are made using industrial waste. However, the project is not just about using local materials, it is about enabling **self-organsation**. The initial architect, London-based Assemble, decided that rather than hand the project on to Belgium based BC Architects they would complete it together as a collaboration. The project is also being treated as a template for other projects, enabling other communities to work collaboratively to mine local waste materials and develop new local vernaculars.

4.2.5 Conclusion on writing a design brief

In this section we have introduced four principles for writing or interpreting a design brief in a way that helps us transition towards more regenerative practice in the construction industry:

- **Aiming higher in the system** prompts us to see each project as a catalyst for wider systemic change
- **Aim to re-establish connection** requires us to redefine the site boundary, to connect the design with the community and its ecosystem, and to establish a long-term relationship with the place and people we are working with
- **Building symbiosis** encourages us to reimagine our supply of resources, creating mutually beneficial relationships and unlocking community capacity
- **Enhancing the capacity to self-organise** encourages us to tune in empathetically to the existing patterns of the dynamic and evolving system in the place we are designing

In summary, how we write and interpret a design brief is an opportunity to transition to a more collaborative approach rather than exerting control over the places we design in. This approach is closer to how an architect might work with a client, and in proposing this approach to interpreting a brief, we can envisage the structural engineer's relationship with the client evolving as we transition to a more regenerative approach.

In Section 4.3 we look at how we can develop ideas in a way that promotes transition towards regenerative practice.

4.3 DEVELOPING IDEAS

Let's recap the chapter so far. Through their many roles in the design process, structural engineers can help enable the built environment industry transition to more regenerative ways of working. Using our Living Systems Blueprint as a guide, we are looking at four stages of the design process in turn to see how each can help us create this transition. So far, we have looked at observation, and brief writing and interpretation. Now we will turn to developing ideas in response to a brief.

Let's start by thinking about how a regenerative approach to idea development might be different to more conventional creative approaches in engineering. Then we can think about how to make a transition towards this more regenerative approach. The main difference comes from the goal of enabling human and living systems to co-evolve. In other words, we are aiming to build the creative capacity of humans and living systems to respond to change **together**.

Once again, we return to the idea of seeing design as a continuation of the existing dynamic processes happening in an ecosystem and in a community. A fully regenerative construction industry would weave our built environment into this living web and empower communities to live and adapt as part of this wider living world.

In this framing, we see idea generation much more as an evolution of the patterns that exist in the community and the ecosystem rather than new ideas that are parachuted in from outside.

That isn't to say that new ideas can't be developed from existing patterns — living systems have evolved highly complex structures from combinations of existing structures[4.17]. Nor is it to say that thinking from other places isn't relevant. For example, seeds from one place get transported to another and new variations — ideas — emerge all the time.

The essence is that in a regenerative system, ideas evolve through an interplay of people, the living world, adaptation of existing patterns, addition of new inputs and best fit to the place. In regenerative design we call this 'emergence'.

To help us think about how we can transition to a more emergent way of having ideas, we will turn to a systems model for creativity.

4.3.1 Introducing the creative systems model

The creative systems model, as proposed by creative psychologist Mihaly Csikszentmihalyi[4.18], constitutes three key elements:

- The Creative Individual
- The Culture, representing the pool of existing ideas
- The Society or Audience, evaluating the proposed ideas

These elements are interconnected, forming a triangular system (Figure 4.17).

In this model, the creative individual draws on the pool of existing ideas to have new ideas. This connection between existing ideas and the creative individual forms the right-hand side of the triangle.

An interesting aspect of this model is the concept of idea gate-keeping. When a new idea is proposed, it doesn't immediately become a part of the pool of existing ideas. There usually exists an audience, serving as a gate-keeper, determining whether the idea holds merit. An analogous example could be the 'liking' system on Instagram. Anyone can upload an image — an idea — to the platform, but it's the audience's 'likes' that boost the visibility of the image on other people's feeds.

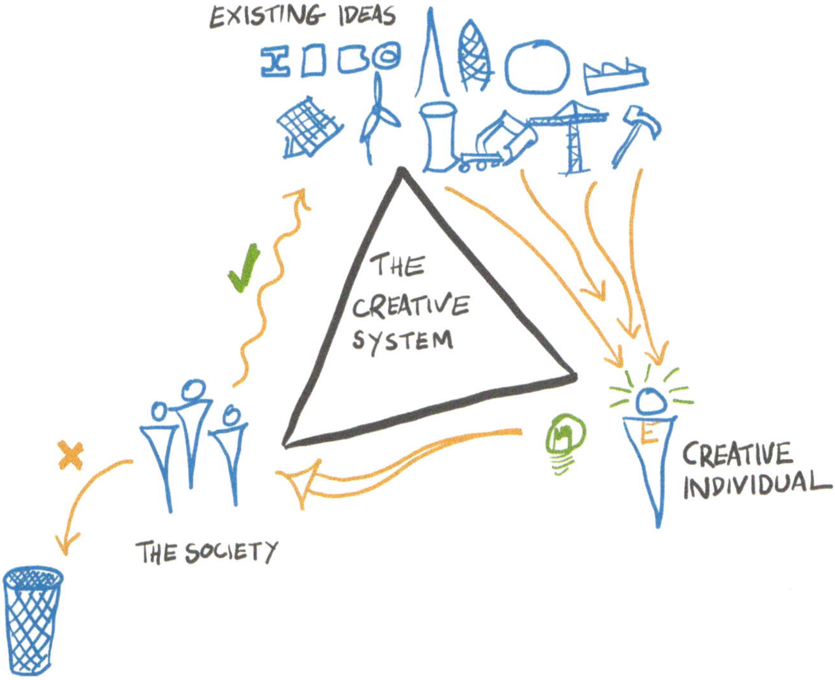

Figure 4.17: Creative systems model proposed by Csikszentmihalyi

So, there is a connection between the person having the idea and the audience that judge it. That's the bottom side of the triangle. And once the idea has been approved — or liked — by the audience, it is transmitted somehow to the pool of existing ideas. That's the left-hand side of the triangle. This connection between the idea creator and the audience forms the base of the triangle. Upon approval or 'liking' by the audience, the idea is then integrated into the pool of existing ideas, completing the left-hand side of the triangle.

Let's look at a simple structural engineering example of this model in practice. The wall of a design office may display various bridge design options. An engineer, upon observing these designs, might incorporate elements of these ideas into a new project. If the internal design team deems these ideas suitable during a meeting, the engineer's sketches may then be pinned to the wall, offering inspiration for the next engineer in need of an idea.

In this example we can identify all the elements and relationships in the creative systems model. The sketches on the wall represent the pool of existing ideas. The engineer is the creative individual and the internal design team is the audience. The connection between the creative individual and the pool of ideas is made when the engineer observes the sketches on the wall. The design team meeting is the connection between the creative individual and the audience. Upon approval by the audience, the engineer adds their idea to the wall, completing the triangular interaction within the model.

The creative systems model helps us connect the creative individual, the pool of existing ideas that they draw from and the process by which new ideas get approved (Case study 18).

CASE STUDY 18

–

The Structural Awards

In 2022 the IStructE's long-established Structural Awards went through a major change. Gone were the prizes for tallest and longest. In its place was a new set of criteria: People; Planet; Process and Profession. As a result, the shortlist of 2022's awards were different to previous years, providing a very different perspective on the best of what engineers can do[4.19]. By making this transition, they showcased a very different set of projects on their website, *The Structural Engineer*[4.20] and into the consciousness of practicing engineers (Figure 4.18).

Figure 4.18: A new approach for IStructE's Structural Awards

4.3.2 The regenerative creative system

In the regenerative creative system, we are interested in asking what would it mean for the communities of people and the living systems they are part of to have ideas together? What might this actually mean? To help us, let's look at each element of the creative system:

- **Who is being creative?** In the context of regenerative design, creative individuals are not just the regular design team, but also community members, the living world, and even the land itself. These individuals offer unique perspectives and insights that can contribute to the creative process, and need to be genuinely brought into the process.
- **What is the pool of ideas?** In regenerative design, the pool of existing ideas is a mix of cultural knowledge, traditional practices, local biodiversity and ecosystem processes. It is made richer by widening the range of cultures included in this pool, bringing global wisdom into a local context. These are all things that we have been immersing ourselves in during our observation phase. It involves not only what has been previously thought up by humans, but also the patterns, cycles and adaptations observed in the local living systems.
- **Who is the audience?** The audience in a regenerative creative system includes both the human community and the rest of the living world in that place. This broadens the idea of 'audience' to involve the beings and entities that are affected by or will interact with the design. The evaluation criteria for a successful idea will thus need to account for ecological resilience, community wellbeing and continued co-evolution.

The connectivity between these elements in the regenerative creative system mirrors the interconnection in living systems. For example, a new idea may emerge from observing an ecosystem process (the pool of ideas), be developed by a local community member (the creative individual), and then be judged successful if it enhances local biodiversity and community wellbeing (the audience).

4.3.3 Transitioning to emergent idea generation

As we have already seen, structural engineers take on many roles to support the design process. In the transition to a more emergent form of idea generation, we can add to this list:

- **As facilitator of the creative process:** enabling a dialogue between place, people and living systems. There is a critical role to play here in empowering all participants in the design process
- **As connector:** making the connections between observations of the existing site, cultural knowledge and local living systems and the design ideas that emerge in conversation
- **As evaluator of ideas:** we can use our skills to help test ideas based on technical understanding and our widening grasp of regenerative systems
- **As educator:** helping to transmit ideas and share observation of the local system in order to widen the pool of potential ideas for creative co-evolution

Through each of these roles, the structural engineer is playing a role in building a part of the regenerative creative system, and helping transition to idea generation that is based on emergence.

The missing piece in all of this is knowing what are the criteria upon which to judge if an idea is a good one. This comes down to testing, which we look at next.

4.4 MAKING AND TESTING

Having made our observations, created a brief and generated ideas, the next stage is to make something at a suitable scale and to test it. This could be a model, a sample panel, a small building (e.g., a pavilion) or used on an existing building.

Testing is the assessing of our ideas against the brief. How would testing work in a design process that was regenerative?

There are two distinct things to think about here:

- **Regenerative tests:** what tests we might apply
- **Regenerative testing:** the process of testing

We will consider these in turn and for each consider how we can transition to a more regenerative approach through the way we apply tests in design.

4.4.1 Regenerative tests

As we set out in Chapter 2, the goal of regenerative design is for human and living systems to survive, thrive and co-evolve. Therefore, the ultimate test of whether a design intervention is regenerative should be measured by whether it **increases** the ability of those human and living systems **within its reach** to **survive**, thrive and co-evolve — and there are three key things to pull out from this sentence:

Increases: The way we have phrased this ultimate test of regenerative design is as a direction of travel. It is not that the design has to lead to a fully regenerative system, but that it **increases** the capacity of the system to regenerate. It is part of the transition.

Within its reach: Recognises that, as we have seen in the Section 4.1, the impacts of our construction activities reach far beyond the site boundary. There is no point in achieving a net increase in biodiversity on site if elsewhere biodiversity has reduced.

Survive: Regenerative design should focus on safety for both people and the environment in the different stages, so all the design standards that make buildings safe are part of regenerative design, as are environmental standards that limit pollution and toxic waste.

In reality, in our current working paradigm, where our supply chains stretch all the way around the world, it is very hard to fully assess the capacity for human and living systems to survive, thrive and co-evolve. But the following two principles can help us transition to a testing regime for regenerative outcomes.

1. **Reduce the reach:** By seeking to localise more of our supply and waste chain, it becomes possible to observe and more fully assess the impacts of our actions. Working at a more regional scale also increases the possibility of interconnection and symbiotic relationships.
2. **Use the Living Systems Blueprint:** This can be a proxy for assessing the human and living system's capacity to survive, thrive and co-evolve. This leads to three tests that are easier to assess:
   - **The interconnection test:** Does the intervention lead to a net increase in connection between the people and the rest of the living world in the system?
   - **The supply test:** Does the intervention lead to a net increase in the capacity for positive symbiotic production of resources?
   - **The co-evolution test:** Does the intervention lead to a net increase in the capacity for human and living systems to co-evolve?

4.4.2 Regenerative testing

The traditional method of design in the construction industry is to establish a brief, generate designs in response to that brief, sign off designs on the basis of that brief, and then proceed to build the building. We have quality processes that check the accuracy of our designs. We have change processes that enable us to modify the design in response, for example, to site conditions or client requirements. But on the whole, in the lifespan of the building and the context of the site, most of the testing of the adequacy of the design happens before the building is built.

In contrast, in regenerative design we view testing as a continuous process. This is because regenerative design acknowledges that in designing for human and living systems we are dealing with complex, interconnected systems that constantly evolve and adapt. As such, the results of our interventions are often unpredictable and will change over time.

In the context of regenerative design, testing transforms into an ongoing, cyclical process. This process continually loops through stages of observing, adjusting the brief, generating new ideas, and evaluating these concepts. It is also a process that can't be entirely known at the start — exactly which tests to apply will only become clear over time. And so, part of our process needs to be generous in allowing for future responsiveness to change.

This idea of continuous improvement is not in itself new in the context of the construction industry — it's embodied in the process known as 'lean construction'. This construction methodology aims to maximise value for stakeholders while minimising waste and improving efficiencies across the board through a mindset of making continuous improvements to practices and methods[4.21]. By analogy, regenerative design could be seen as a lean process — one that is rooted in achieving best alignment with human and living systems, enabling them to survive, thrive and co-evolve.

The concept of this iterative, long-term approach isn't new either. There are Indigenous cultures that advocate for designing with the foresight of seven generations into the future[4.22]. This idea is encapsulated in the Welsh Well-being of Future Generations Act[4.5], which obliges public bodies to consider the long-term impact of the decisions they make, and which has led to a significant reduction in road building in Wales[4.23].

To transition towards a more continuous testing approach, we propose the continuous, place-based design process, described in Section 4.5.

In summary, regenerative design transforms testing from a one-time, gatekeeping activity into an on-going, iterative and long-term process. In doing so it aligns more closely with the living world's complexity, interconnection and capacity for adaptation and evolution. Regenerative design requires us to take a much more longitudinal relationship to our designs and the places we are designing in. But by doing so, we can create designs that are a more effective extension of the living systems that we are part of.

4.5 CONTINUOUS, PLACE-BASED DESIGN

To encapsulate the ideas in this chapter, we propose a new diagram for the design process. We call it continuous, place-based design (Figure 4.19). The diagram emphasises:

- The importance of establishing a long-term relationship with place
- A design process that begins with observation
- An approach that works iteratively with the complex ecological and social systems around us in order to find the design that best realises the potential of these systems

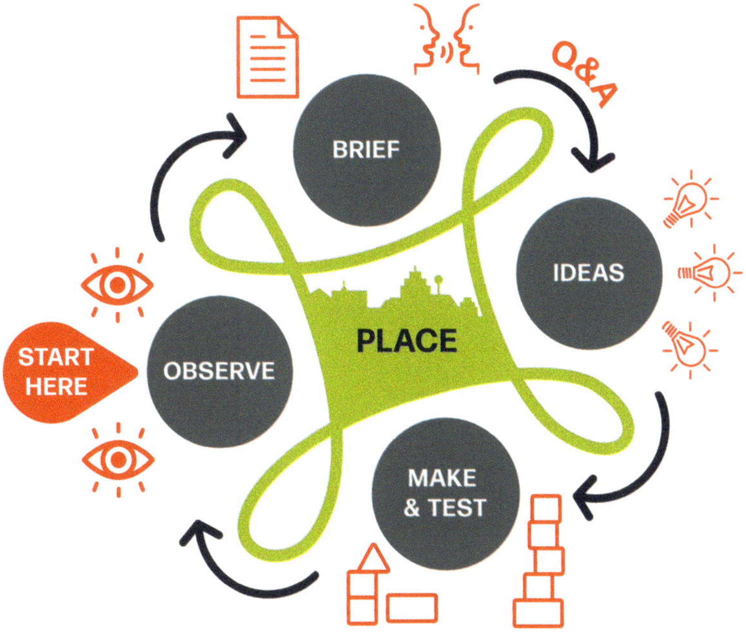

Figure 4.19: Continuous, place-based design

By seeking to adopt the philosophy of continuous, place-based design in our projects, we can help to build a transition to a more regenerative industry through the way we design.

4.6 CONCLUSION

At the beginning of this chapter we asked: **How can we use design to help us transition to a more regenerative construction industry?** To answer the question, we defined four stages in a design process and looked at how we would conduct these regeneratively. These stages are observation, brief writing, idea development, and making and testing. Each stage provides opportunities for structural engineers to help build a transition towards a more regenerative construction industry and Table 4.1 provides a summary of key points.

Table 4.1: Summary of key points from this chapter (Chapter 4)

Key stages in design	Transition steps
Observation	Extended research period going beyond traditional desk studies to form deep connection with place. We consider four observational lenses.
Ecosystem	Identify what lives there, understand life-enabling factors, what is missing that would enable life to flourish, study topography and patterns of life, look for symbiotic opportunities, recognise existing living cycles and their influence.
Materials	Understand origins and impact of materials, consider rate of material renewal, look for options to repurpose existing materials, consider material transport impacts, study waste streams, and build relationships with suppliers.
Community	Identify individuals affected by the project, understand local connectivity, resilience and relationship with the living world, investigate inequality factors, recognise community abundance in culture, heritage, knowledge and skills, tap into local knowledge and trades, consider potential impact on future generations.
Local construction	Understand vernacular design and its implications, evaluate state of existing buildings, study traditional materials and their properties, investigate construction techniques unique to the location, identify lost or potential revival of construction methods.
Brief	The point of greatest opportunity to set goals and parameters that help to transition towards regenerative practice. We introduce four principles:
1. Aim higher in the system	Extend vision beyond project boundaries, see projects as catalysts for systemic change, think holistically, influence broader systems like supply chain organisation, local planning laws, governing codes of practice, drive research and innovation, shape organisational culture, and transform shared values.
2. Aim to build interconnection	Acknowledge that projects don't exist in isolation, redefine traditional parameters, expand the 'site' to include impacted places ('second site' concept), connect with ecology and community, foster relationships of stewardship between community and ecosystem, form enduring relationships with the site and its community, ensure long-term engagement.

Table 4.1: *continued*

Key stages in design	Transition steps
3. Aim to build symbiosis	Shift focus towards resource supply, move from consumption to production, create conditions for symbiosis, restore abundance in depleted systems, leverage regional abundance in materials and skills, unlock community capacity.
4. Aim to enhance capacity to self-organise	Look for the unique potential of a place and how that can be unlocked. Ask questions about repeating patterns on the site, enabling factors, and the most appropriate evolution. This should be done through four lenses: society, living things, buildings and transport.
Idea generation	Building capacity for ideas to emerge from the community and the ecosystem. Use the creative systems model to guide thinking.
Build the creative system	The creative individuals in regenerative design include not only human designers but also local communities, living world, and even the land itself. The pool of existing ideas draws upon cultural knowledge, traditional practices, local biodiversity, and ecosystem processes. The audience, composed of both human and non-human entities, judges the success of an idea based on ecological resilience, community wellbeing, and continued co-evolution.
Transition to emergent idea generation	Structural engineers play multiple roles in transitioning to emergent idea generation: as facilitators of the creative dialogue, connectors linking observations with emerging ideas, evaluators employing their technical skills, and educators transmitting knowledge to broaden the pool of ideas.
Making and testing	Is it enabling a transition to regenerative practice? We consider what tests to apply and how to apply them.
What we test for	The test of whether a design intervention is regenerative should be measured by whether it **increases** the ability of those human and living systems **within its reach** to survive, thrive and co-evolve. Two things can help: • **Reducing the reach:** Localise more of our supply and waste chain to more easily assess the impacts of our actions • **Using the Living Systems Blueprint** as a proxy and test for interconnection, symbiotic supply and self-organising potential
How we test	We are dealing with complex, interconnected systems that constantly evolve and adapt. As such, the results of our interventions are often unpredictable and will change over time. To transition towards a more continuous testing approach, we propose the continuous, place-based design process.

The continuous, place-based design diagram (Fig. 4.19) helps us to envisage what a regenerative design process could look like, and gives us clues about the things we need to change in the way we work to create the transition we are looking for.

Of course, while you've been reading this chapter, you've probably been thinking of the barriers to doing these things in practice. Going back to our Systems Bookcase model, we are reminded that lots of things that control what we can and can't do in design are set higher in the system, at the level of operations. It's almost time now to go one shelf up on the systems books case and ask: **What can the regenerative structural engineer do to shift the built environment towards more regenerative operations?**

But before that, we present two, more detailed case studies.

CASE STUDIES 19 AND 20

Throughout this chapter we have included a number of short case studies to highlight specific aspects of the continuous, place-based, design which can be utilised to move towards regenerative outcomes. However, here we present two longer case studies that capture a large number of different aspects in a single practice or project.

CASE STUDY 19

—

Using felled urban street trees, UK

London-based Yes Make specialise in working with felled street trees in London and turning these into community-structures and infrastructure, working with people from the local community to do the design and build.

Their work began in 2021 when founders Joel De Mowbray and Morgan Da Silva observed that 5-10,000 trees come down in London every year, the majority of this wood turned into woodchip for incineration, and came up with the brief:

> "what if the tree that grew in your street could furnish your home, could create your child's playground and build the community spaces that bring us all together?"

Today, the organisation they founded, Yes Make, is a hub for sourcing locally felled timber and other reclaimed materials and integrating these into local community projects. When Oliver went to visit them at their operating site in Canada Water, a group of young people were working on the largest cordwood building to be built in the UK. The primary structure is constructed from reclaimed timber sleepers donated by National Rail, while the timber for the cordwood infill panels came from trees recently felled in Epping Forest. Yes Make have converted an old milk float into a flatbed vehicle complete with hoist for transporting their materials around the city (Figure 4.20).

Figure 4.20: Converted milk float transporting one of the thousands of street trees

Yes Make are best known for creating the Ark, a sheltered space to enable the community of Hither Green to meet outdoors while maintaining social distancing during the Covid-19 pandemic (Figure 4.21). It is recognisable for its Y-shaped structural supports, each uniquely formed from branching trees. Branching timber is usually of low value because it is irregular. But here the special shape of this piece becomes a unique part of the structure.

As Morgan describes, structural engineers working in this way need to be able to do two things. The first is to engage with materials on a project-by-project basis. Instead of knowing that you have a regular material to work with, the material is different in every project. Even if it is something a bit more standard, like reclaimed steel, you still have to check it over. A key skill here is having good materials literacy, not just of standard products but when engaging with a much wider range of materials. Where are the strengths and weaknesses? And even if something might be weaker than we are used to, can it still be useful?

Figure 4.21: The Ark, made from the Y of different street trees

The second requirement is creativity. To start with a set of existing materials and ask: What can we create with these? We see this as flipping the story: usually structural engineers start with an idea and know that they have a standard palette of materials to build from.

Yes Make are engaged in continuous, placed-based design, finding new ways to work with materials, and figuring out how they can make their innovative ideas work within the limitations of the existing building code.

An important part of Yes Make's work is community involvement in design and construction. They run workshops with people in their community to build the local material literacy that they see as so important. Getting involved with design and construction builds local experience, skills, cohesion and a sense of ownership over their projects. And Morgan says, there is a strong emotional attachment to trees. When people see timber from local trees being used to create a local structure, we are continuing that sense of attachment and connection.

A question that often comes up when we discuss Yes Make's work with engineers is: What happens when the supply of felled trees runs out? We see this question as coming from a place of recognising that the UK construction industry uses much more timber than can be supplied by end-of-life urban trees. But Yes Make's response is refreshing. As Morgan describes, when the supply of trees has run out, they will have achieved what they wanted to, they will have answered the brief, and they will move on to new observations to find what other waste streams can be turned into an input.

Their vision for growth is a mesh of organisations working creatively with each other to exchange materials, skills and ideas that meet the needs of the community. We see this approach as rooted not in the scarcity-minded frame of what happens when things run out but rather in the abundant-minded frame of what could we create together.

To conclude, from a regenerative design perspective, Yes Make are:

- Using a continuous process over many projects to see how urban trees can be integrated into a circular supply chain
- Working with an abundant renewable resource in their ecosystem and finding ways to bring that resource into use
- Creating a link between construction materials, where they are sourced and the people that use them
- Building local capacity to self-organise by developing design and construction skills (including material literacy)
- Operating at a community scale, seeking to prioritise reciprocity and connection rather than growth

CASE STUDY 20

Ilima Primary School, DRC

The Ilima Primary School was built in the Equator Province of the Democratic Republic of the Congo. The design, carried out by MASS Design Group, took account of the local constraints where there was incredibly limited site access (multiple single log bridges and narrow paths) which limited what materials could be imported to site (Figure 4.22).

Figure 4.22: Cement being transported by motorbike

As a result the design and construction focussed on maximising the use of local materials and craft. This led to loadbearing earth-block walls utilising soils taken from termite mounds mixed with clay. The decision to use termite mounds was based on local knowledge which was acquired by listening to the community.

The roof was constructed from timber, which was locally sourced and was a similar species to cedar. A different locally sourced timber was turned into shingles and used to construct the roof. This was used instead of corrugated steel sheet, which would have needed to be transported and carried across the bridge by hand, and would be costly and time consuming to replace over the life of the building (Figure 4.23).

Figure 4.23: Ilima Primary School. An investment in conservation education at a critical intersection of human and ecological habitat

Testing was carried out to characterise material behaviour of both the earth and timber. The timber shakes naturally degrade over time and need replacing — this is a good thing as it keeps the local knowledge alive!

The building was constructed using local labour (748 equivalent full-time year-long jobs were created), upskilling local residents (120 local labourers were trained in Lo-Fab construction). Due to the local sourcing of the materials 93% of the project budget was spent locally, and 83% was spent on labour. The project team calculated an embodied carbon (A1-A3) of just $15kgCO_2e/m^2$ which is well below the global average for education projects. The project resulted in six classrooms which aimed to increase pupil enrolment by 40%, but the actual increase was given as 200% by the head teacher. There have been two new community agreements (adoption of land use plan and refrain from eating bush meat) and 99% of materials were sourced from within 10km of the building.

While you could argue that the design is sustainable predominantly because of its location, both globally and locally, and that it would be hard to replicate this in another setting, we would argue that the design responds to its local constraints, achieves significant positive social impact, while minimising environmental impact because of the willingness of the designer and community to collaborate, look past standard solutions towards community inspired solutions, and through testing, was able to utilise local materials not normally considered. While the specific solution is unlikely to be adaptable to your local setting, the general principles around the approach to regionally abundant materials and labour, are.

REFERENCES

4.1 The Institution of Structural Engineers. *Conceptual design of buildings*. London: IStructE Ltd, 2020.

4.2 Wilson E. 'Disabled perspectives in nature writing'. *It's Freezing in LA!* 2022:(9). Available at: https://itsfreezinginla.com/articles/nature-reframed [Accessed: December 2023].

4.3 Hebel DE et al [eds]. *Addis Ababa: A Manifesto on African Progress*. Berlin: Ruby Press, 2019.

4.4 Hildyard D. *The Second Body*. London: Fitzcarraldo Editions, 2017.

4.5 Welsh Government. *The Well-being of Future Generations*. Available at: www.gov.wales/well-being-of-future-generations-wales. [Accessed: December 2023].

4.6 YouTube. [TED] Yasmeen Lari: *No to handouts. Yes to self-reliance*. Available at: www.youtube.com/watch?v=1LDPIvM0Qcg [Accessed: December 2023].

4.7 The Guardian. *Interview: Architect Yasmeen Lari: 'The international colonial charity model will never work'*. Available at: www.theguardian.com/artanddesign/2023/may/07/architect-yasmeen-lari-royal-gold-medal-2023-interview-pakistan-barefoot-social-architecture-floods-earthquake-what-a-terror-i-was-when-i-designed-those-mammoth-buildings [Accessed: December 2023].

4.8 Heyne Tillet Steel. *Steel Reuse: Incorporating reclaimed steel at Holbein Gardens*. Available at: https://hts.uk.com/blog/reuse-of-steel/ [Accessed: December 2023].

4.9 The Institution of Structural Engineers. 'Holbein Gardens: delivering a low-carbon structure with reclaimed steel'. *The Structural Engineer*. 2023:101(3). Available at: https://doi.org/10.56330/CRPP8446 [Accessed: December 2023].

4.10 Heyne Tillet Steel. *Holbein Gardens*. Available at: https://hts.uk.com/projects/holbein-gardens/ [Accessed: December 2023].

4.11 Heyne Tillet Steel. *Steel Warriors — Calisthenic street gym*. Available at: https://hts.uk.com/projects/steel-warriors-calisthenic-street-gym/ [Accessed: December 2023].

4.12 Layard R and De Neve J-E. *Wellbeing: Science and Policy*. Cambridge: Cambridge University Press, 2023.

4.13 Nationale Bruggenbank. *Welcome to the website of the National Bridge Bank.* Available at: www.nationalebruggenbank.nl/en/ [Accessed: December 2023].
4.14 WeCanMake. *WeCanMake is a neighbourhood test-space in Bristol.* Available at: https://wecanmake.org/ [Accessed: December 2023].
4.15 WeCanMake. *WeCanMake Playbook.* Available at: https://wecanmake.org/wp-content/uploads/2023/03/wecanmake_playbook.pdf [Accessed: December 2023].
4.16 The Guardian. *The ultimate eco building — made of salt, sunflowers and recycled urine.* Available at: www.theguardian.com/artanddesign/2023/jun/05/ultimate-eco-building-salt-sunflowers-recycled-urine-atelier-luma-architecture [Accessed: December 2023].
4.17 Kennedy J. *Pathogenesis: How Germs Made History.* London: Penguin, 2023.
4.18 Csikszentmihalyi M. 'Implications of a Systems Perspective for the Study of Creativity' in *Handbook of Creativity* by Sternberg RJ [Ed.]. Cambridge: Cambridge University Press, 1998.
4.19 The Institution of Structural Engineers. *Winners of the 2022 Structural Awards announced.* Available at: www.istructe.org/resources/news/winners-of-the-2022-structural-awards-announced/ [Accessed: December 2023].
4.20 The Institution of Structural Engineers. 'Complete issue (November/December 2022)'. *The Structural Engineer.* 2022:100(11). Available at: https://doi.org/10.56330/YOEV2807 [Accessed: December 2023].
4.21 Lean Construction Institute. *What is Lean Thinking?* Available at: https://leanconstruction.org/lean-topics/what-is-lean-thinking/ [Accessed: December 2023].
4.22 Kimmerer RW. *Braiding Sweetgrass: Indigenous Wisdom, Scientific Knowledge and the Teachings of Plants.* London: Penguin, 2020.
4.23 UBM Built Environment '"World-leading" net zero policy sees Wales cancel road building projects'. *Building.* Available at: www.building.co.uk/news/world-leading-net-zero-policy-sees-wales-cancel-road-building-projects/5121784.article. [Accessed: December 2023].

CHAPTER 5

Creating transition through operations

How can a structural engineer facilitate systemic change at the operations level?

In Chapter 2 we described the construction industry as a system of various components — elements, buffers, feedback loops, information, rules, self-organisation, goals, and the overarching paradigm. We saw how regenerative design is a new goal for our industry, aligned with a holistic paradigm.

In Chapter 3 we used the Three Horizons Model to show that while regenerative design may be something many in our sector are committing to in principle, it is not something we can fully comprehend within our current way of working. Hence, we defined the 'Living Systems Blueprint for Regenerative Design' to navigate our way. The blueprint highlights key characteristics of thriving, living systems: interconnection, symbiosis, and capacity to self-organise and evolve. We said that the role of the regenerative structural engineer is to use this Living Systems Blueprint to guide the transition towards a more regenerative construction industry.

In Chapter 4, we delved into the first tier of the Systems Bookcase model — design. We asked "How can design serve as a vehicle for transition?" We then explored how design practices, when guided by the Living Systems Blueprint, can lead us towards a more regenerative construction industry.

But there is, of course, a limit to how differently a structural engineer can act in their design work — a limit to how far they can stretch what is permitted in what they design, and that is to do with the rules of the system they work in. If we think back to our Systems Bookcase model, what is allowed on the design shelf is defined by the **information, rules and regulations** on the **operations** shelf above. So if we want to shift design towards more regenerative practice, we need to shift the operating rules of our industry towards more regenerative practice.

So, in this chapter, we now move up to the operations shelf of the Systems Bookcase (Figure 5.1). Here, we explore the interactions and relationships that shape our industry at a broader level.

Given the complex, interconnected nature of these various realms, we will avoid an overly compartmentalised approach. Instead, we will use four case studies to demonstrate how these different areas can intersect and inform one another. Each case study will illuminate how different aspects of the operations level can contribute to the transition towards a more regenerative construction industry.

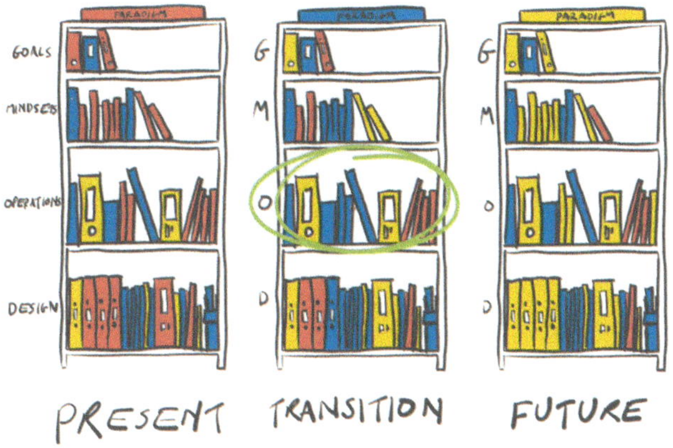

Figure 5.1: Changing the operations shelf in the system

5.1 OPERATIONAL CONTEXTS

Before we delve into the different information and rules that will change the operations shelf of the bookcase, we define five operational contexts that engineers work in. Each of these have their own information flows, rules and cultures. As a result, engineers can influence each of them in different ways.

1. **Place of work**: This is perhaps the operational context with which we most easily associate. It is where we bring our skills and experience, share our ideas, review others' ideas and collaborate to enable projects to happen in exchange for a salary. In this context, at a systems operations level, we can ask:
    - What do we see as our place of work? An office, a community, an ecosystem? A forester goes to work in a forest. There is a clear connection between where they work and the living system they are working with. Like a forester, our work impacts living systems, but our places of work are separated from the living systems we impact. So how can we reinforce this connection through our places of work?
    - Who are we working with? Do we have a shared goal of achieving regenerative outcomes and do we support each other in this?
    - What jobs do we work on? Do we have a bid/no bid policy that helps select the most suitable projects? Can we revise these to move towards regenerative projects?
    - What review processes do we have? Can we create a set of questions that encourage regenerative practice?
    - What projects are shared and celebrated in the office? Can we utilise these to share best practice?
    - What culture do we have? Have we created a culture than values health and safety and regenerative design above all else?

Figure 5.2: Opportunities for sharing regenerative practice in the office

- How is the office set up? Do we have pictures on the wall that inspire regenerative design? What books are in the library and what magazines do we include? Is there an abundance of indoor plants? Do we challenge hierarchies and make the office inclusive? (Figure 5.2).

(We explore the influence we can have in our place of work in more depth in Chapter 6, Section 6.3).

2. **Collaborators in industry**: These are the individuals and organisations we work with across industry to help us deliver our work. In this context, at a systems operations level, we can ask:
   - Who are the people we need to be working with to create greater interconnection with place, community and ecology?
   - What are the opportunities and barriers for mutually symbiotic working relationships?
   - How do ideas emerge in these collaborations? How do we build capacity into the relationships? How do we sustain and grow these relationships?

3. **Supply chain dynamics**: This relates to the set of processes and the people involved that provide the resources that we work with. In this context, at a systems operations level, we can ask:
   - How are we building connection between ourselves, the places materials come from and the communities we are designing for? Probably the biggest shift we can expect in a transition to a regenerative construction industry is a closing of the gaps between designers, communities, and the places materials come from
   - How are we transitioning away from extractive materials supply towards materials drawn from abundant resources and mutually symbiotic living systems?
   - How can we unlock experimentation in the supply chain to allow use of more local and/or non-standard materials?
   - How can we enable supply chains which are operating in a regenerative way?
4. **Civic sphere**: This is the way in which construction relates to the governance of the communities, towns, cities and regions in which we work. In this context, at a systems operations level, we can ask:
   - How can engineers play a greater role in the planning process to help us transition towards more regenerative outcomes?
   - How can we create a more shared sense of ownership of our living commons: air, water, soil and biodiversity?
   - How can we transition towards a design process that is community-led and community-empowering, drawing on local knowledge and building local capacity to respond to change?

5. **Wider industry**: The network of organisations and institutions that drive innovation and regulate practice. In this context, at a systems operations level, we can ask:
   - How can guidance on professional practice positively support and grow the remit of the structural engineer to consider a wider range of factors in design?
   - How can we use and share the abundance of data, processing power and AI at our disposal to unlock new ways of designing more regeneratively?
   - How can we create industry regulation that enables experimentation with emergent ideas rather than locking in methods of working that may no longer be fit for purpose?

Taking all of this into consideration, our overarching aim is to answer the question: ***How can a structural engineer facilitate systemic change at the operations level in the construction industry?***

5.2 CASE STUDIES

Perhaps the best way to answer this question is via case studies. The four in this section demonstrate how we can move up the Systems Bookcase as we transition towords a regenerative future. Each begins by describing the regenerative approaches at the design level, and then explains how the practitioners aimed higher in the system — how they looked to change the operations shelf in a variety of ways.

CASE STUDY 21

-

Rwanda Institute for Conservation Agriculture, Rwanda

Our first case study provides an insight into how an interdisciplinary design team leveraged an abundant local material, effectively broadening the opportunity for its use on a much larger scale. Their approach emphasises the power of interdisciplinary design teams to unlock the potential of local resources.

The Rwanda Institute for Conservation Agriculture (RICA) aims to enable the education of Rwandans in sustainable food production. It includes accommodation, teaching spaces and workshops and in many ways follows the typology of a university (Figure 5.3). RICA is the initiative of the Howard G Buffett Foundation. MASS Design Group worked with them from the beginning to develop the vision, programme, and budget for the Institute. For MASS the project was an opportunity to explore what is possible when approaching design with a regenerative mindset.

Figure 5.3: The campus, built around a One Health approach, trains the next generation of leaders in conservation agriculture

RICA itself encompasses many attributes that help human and living systems survive, thrive and co-evolve:

- It aims to preserve and restore the local ecosystem
- It aims to embed students in the local community and ecosystem, helping to establish interconnection
- It aims to create places for community and enterprise using local materials and labour — unlocking the symbiotic potential of place. Similarly, the project uses furniture locally made from local materials

We strongly encourage readers to explore the comprehensive scope of the overall project. However, for this book, we've concentrated on the structural aspects, highlighting actionable steps and opportunities for engineers within their current professional framework.

What we find so compelling about this project is MASS's innovative thinking. By taking a multi-disciplinary approach and removing traditional boundaries around design disciplines they sought to release local potential and abundance, effectively acting as opportunity seekers, not just problem solvers.

Project details

Let's look at some of the details of the project. RICA includes 69 buildings with a combined footprint of 12,750m^2.

The project aimed to source labour locally, with 98% of labour coming from within 100 miles of the site, and a large proportion of the labour coming from the local vicinity (Figure 5.4). This strategy turns the construction process into a local catalyst, knitting together the community with the places they are shaping, while simultaneously bolstering local capacity and enriching collective knowledge.

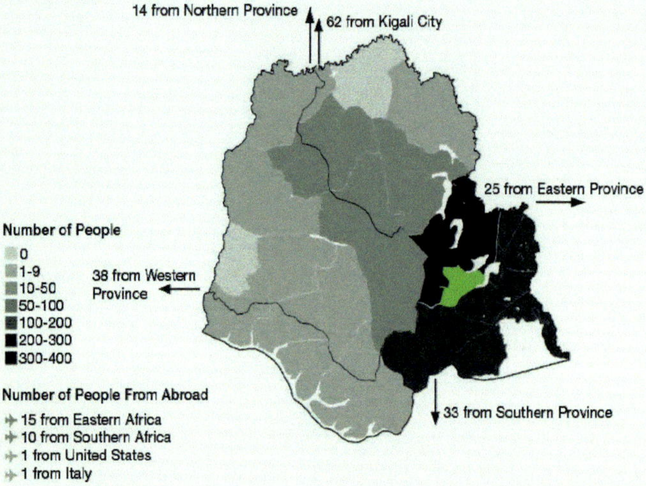

Figure 5.4: Geographical distribution of labour

The project sourced 96% of its materials from within Rwanda, and the structure included:

- Stone foundations
- Wood framed roof structures
- Rammed earth and non-fired compressed earth block walls (Figure 5.5)
- Coffee-husk fired ceramic tile roofing (Figure 5.6)

Figure 5.5: Rammed earth walls. The workers were trained by global experts in rammed earth construction and have gone on to form a cooperative which continues to build rammed earth walls in Rwanda

Figure 5.6: Natural materials are used throughout, including coffee husk fired clay tiles and stone cobblestones

An example of how MASS moved away from degenerative practice and towards regenerative practice is the design of the roof trusses. In Rwanda trusses are normally imported steel, which utilise a distant workforce and a highly emissive material. Instead, MASS chose wood trusses to promote a locally available, abundant material that stores biogenic carbon and is manufactured by local people. There are no service life implications, a wood building can last longer than steel buildings if designed and detailed correctly in this environment. In this case there were no useage implications because MASS designed the space to be within the material's capacity. From a construction cost perspective there was no discernible difference between steel and wood trusses (potentially the wood trusses were cheaper). The reason wood trusses are less common is due to the young supply chains and lack of structural grading, which are a real barrier.

This approach not only significantly reduced the extent of the supply chain, but it also fostered enhanced **interconnectivity**, strengthening the bond between the designers and the origins of the materials. Furthermore, by prioritising materials that were either abundant and/or renewable, the design shows a commitment to building a material supply chain that interacts positively with the places it touches.

The project included a detailed analysis of the supply chain and the carbon emissions in the supply chain, based on local knowledge where possible. As a result, the carbon footprint was reduced from 507kgCO_2e/m^2 using a "business as usual design" to 214kgCO_2e/m^2 with the final design. This demonstrates the principle of emergence: working with local knowledge to find a design with better fit to the local conditions.

The project makes significant use of timber, and during the early stages of design the New Forest Company in the south of Rwanda achieved the first FSC certification in the country, so the team were hoping to use them exclusively. However, there was insufficient supply, so MASS travelled to mills in Rwanda, Uganda, Tanzania and South Africa to get a better understanding of the capacities, practices and processes of several mills.

It was decided that the other main source of wood should be from Sao Hill in Tanzania, which does not have FSC certification. They did, however, have a Forest Management Plan that certainly addresses key issues of biodiversity and reforestation, and they were aiming to be FSC certified by the end of 2023.

All of these facets effectively illustrate many of the practices detailed in Chapter 4, providing a real-world demonstration of how design can catalyse the transition towards a more regenerative construction industry.

Aiming higher in the system

Beyond the individual project, we also want to demonstrate how this scheme worked higher up in the system, changing the operations shelf by providing information, creating new rules, and enabling new industries to emerge.

To enable the construction of rammed earth and compressed block construction, MASS required a suitable contractor. The problem was that at that time, a rammed earth and compressed block contractor didn't exist in Rwanda so they set about creating the conditions that would enable a contractor to emerge. MASS worked with the project's general contractor to locate external expertise and bring them to Rwanda to oversee the training and construction of the project's various earth walling (Figure 5.7). On completion, the team trained on earth construction established an independent, Rwandan-owned and operated business, continuing to contribute to the local industry after the project had finished. This example demonstrates how engineers can strategically intervene at the level of the supply chain, effecting changes that reshape local industrial operations and open new avenues of practice.

Figure 5.7: Construction of pressed block earth walls

One of the major obstacles to the use of earth construction was the absence of appropriate design codes and regulations for rammed earth in Rwanda. In response, MASS worked with the regulatory authorities to create a Rwandan design code for adobe earth block construction (90% of Rwanda's population live in adobe block buildings). This initiative not only set the guidelines enabling others to design with this material within Rwanda, but also provided a valuable reference for designs beyond the country's borders[5.2].

MASS also challenged structural engineers globally to consider stone foundations as a viable option. The project was featured in a 2021 issues of *The Structural Engineer*, even making the front cover (Figure 5.8) This detailed article, outlining how the use of stone foundation is possible, effectively broadened the horizons of their fellow engineers and themselves — MASS went on to use stone foundations on their "Stone Barns" project in the US.

Figure 5.8: Stone foundations on the cover of *The Structural Engineer*

In taking these steps, MASS has instigated changes within the wider industry. They've helped to shape rules and regulations that expanded the acceptance and usage of locally abundant materials. This story illustrates how engineers can exert a meaningful influence on industry practices, driving the adoption of more locally-appropriate and sustainable material choices.

In addition, the project had a profound impact on a number of the designers working on it. Will Arnold has gone on to lead the IStructE's response to the climate emergency. Kelly Alvarez Doran has gone on to lecture at the University of Toronto where he is challenging the paradigms of design and the unquestioned norms to suggest new approaches and rules based on what Canada and USA can learn from Rwandan design approaches. Other designers, such as James Kitchin talk about how his paradigm for design has changed, how working on this project has changed his approach to design. And the project has been shared a number of times to challenge what engineers consider possible, not least, here in this book!

In summary, MASS Design Group's work on RICA is a fantastic example of the power of design teams to challenge industry norms, instigate wider changes and inspire regenerative behaviours. In their work, they intervened higher in the system to create change that can help transition construction industry operations towards more regenerative practice.

CASE STUDY 22

-

Modcell®, UK/New Zealand

Our second case study delves into a similar theme — scaling up the use of an abundant material. However, this time, instead of focusing on a single client's needs, the design team identified a broader industry need and developed an innovative solution to meet it. Their visionary thinking demonstrates how designers can contribute to transforming industrial practices and standards, pushing what is possible and desirable.

Modcell® is a product that emerged from the aspiration to make greater use of straw, a bio-based waste product, in the built environment[5.3]. The concept was to integrate this abundantly available resource with off-site fabrication techniques, thereby making it easier to design with and specify.

James has a direct connection to Modcell®. One of the reasons he joined Integral Engineering Design was to work with Margaret Cooke and Tim Mander, who were directors of Modcell®, along with Craig White and Linda Farrow of White Design, who went on to found Agile Homes.

Figure 5.9: Completed in 2002 the first Modcell® panels can be seen, installed in the end wall of this building at the University of the West of England

The first use of Modcell® was for a wall at the University of the West of England, but the company went on to supply several other projects (Figure 5.9). What interests us here is how the development of Modcell® exemplifies continuous, place-based design, and how they were able to think beyond projects to make change at a more operational level in the construction industry system.

Project details

In Modcell®, the designers took an abundant material, waste straw, and converted it into prefabricated wall panels with high insulation properties and good breathability. The panels used timber to create a frame and were finished with boarding or lime render. On many projects the straw was locally sourced, and panels were made in a flying factory (a factory set up just for the duration of the project) which was located as near to the construction site as possible. Labour was locally sourced (and sometimes included the community).

Modcell® has been used to create some amazing buildings, three of which we'd like to highlight here.

The first is Knowle West Media Centre in Bristol (also mentioned in Chapter 4 as part of the WeCanMake case study). The Media Centre was one of the first major Modcell® projects (completed in 2008, which serves as a reminder of just how long building with renewable/low carbon materials has been a priority for some firms), and remains an essential part of the community; acting as a home for a number of community focused projects (Figure 5.10).

Figure 5.10: Knowle West Media Centre

The second project is the science block at Hayesfield Girls' School in Bath, which is constructed from Modcell® wall and roof panels. It has moved away from the traditional vernacular of Modcell® buildings, with a lime render applied across the full face and a DNA helix etched into the render (Figure 5.11). It has a low embodied carbon (and, could be said to sequester carbon). It also has low operational energy requirements of 17.5kg $CO_2e/m^2/y$ and it was constructed using a flying factory — with some of the wall panels constructed by the students. While it is hard to define "most sustainable building" this is the building James always thinks of when discussing sustainability[5.4].

The third project is LILAC (Low Impact Living Affordable Community) in Leeds. LILAC is a co-housing

Figure 5.11: 'The Nucleus' at Hayesfield Girls' School

development with many regenerative features. Modcell® was a natural fit, but the project is also regenerative in a number of different ways. Many of these can be read about in *Unlocking Sustainable Cities* by Paul Chatterton[5.5], who is Professor of Urban Futures, and also was part of the LILAC project and currently lives there. In this example, engineers and architects have enabled the project to happen, but also Modcell®, which existed outside of the project, has enabled it to be as low impact and biobased as possible (Figure 5.12).

Figure 5.12: LILAC – ecobuild households in West Leeds

Already, we can see that the approach taken by the designers in creating Modcell® is a great example of how, through design, engineers shift the construction towards more regenerative practices. It is also an example of applying the continuous place-based design process. Their strategy includes using renewable resources that are abundant and easy to source. They focus on connecting the materials they use to the place they're used in, making their projects more integrated with the local environment. On top of that, they work on developing local skills and building capacity, helping the community grow with the project. These are great illustrations of some of the ideas developed in Chapter 4. But again, what interests us here is how the work developing Modcell® illustrates structural engineers intervening higher in the system to change the way the construction industry operates.

Aiming higher in the system

The creators of Modcell® were motivated to harness the benefits of straw-bale construction while overcoming its rustic image, transforming it into a scalable construction material that could be used in a range of contexts. Until then, straw-bale design had been seen as rustic, hard to replicate and difficult to certify. What they wanted to create was a replicable, growable system that could be adapted to the needs of the local area and the project. The development of Modcell® led to a large amount of research in collaboration with the University of Bath, which has subsequently been published.

This is an example of structural engineers working with their *collaborators in industry* not on a single project but aiming to intervene higher in the system, to create a solution for industry, not just for a single client. This was about spotting an opportunity to leverage modern methods of construction to overcome historic obstacles, and creating a product ready to serve projects even before they were conceived.

Through collaboration between architect, engineer and researchers, the team were able to enable the development of the product, they were able to influence the specification (and not just the specification, but what was delivered to site), they were able to advocate, for example for the use of flying factories, which generally were more expensive but offered other benefits. They acted as opportunity seekers, developing and iterating the product over a number of years, to find different ways to utilise the product. And they were able to use their creative talents to produce some truly amazing buildings.

Modcell® and its associated projects won a number of awards over its existence, and while it has currently ceased trading it maintains an active internet presence and still holds the registered trademark.

However, this was not the end of the regenerative project. Craig White and Linda Farrow, two of the directors, and local to Bristol's Knowle West, could see a significant housing need. They took their learning from Modcell® and developed a new straw-based construction method to create homes for the most disadvantaged. One of the biggest barriers to this was the funding of such projects, and so Craig became an expert in community funding and has set up a community saving and investment scheme to enable the funding of new, low cost and highly sustainable houses.

Tim Mander retired in 2021 but remains a strong advocate for sustainable design. Margaret Cooke is still a director at Integral Engineering Design, was one of the co-authors of the IStructE's *Design for Zero*[5.6], was awarded the Mallet Milne Medal in 2021[5.7] and remains active in reuse and historic structures.

Paul Jaquin, an employee of Integral Engineering Design, who worked on much of the research, emigrated to New Zealand. Taking learnings from Modcell®, and other prefabricated straw-bale systems such as Econocon, Paul has worked with a contractor, Hiberna, in the South Island of New Zealand to produce prefabricated straw-bale panels for house construction. The panels are made from locally sourced Douglas fir, which has been sawn at a local sawmill to a bespoke size to allow the most efficient use of materials. The straw was sourced locally from the south part of New Zealand, and was straw which may otherwise have been burned, giving a double saving of reducing carbon introduced into the atmosphere, and locking CO_2 away in the buildings. Construction of the panels is undertaken locally in one of the central construction sites, then the prefabricated panels are transported by loader crane to the required site and lifted straight into place (Figure 5.13).

Figure 5.13: Modular straw-bale construction in New Zealand

This provides for quick construction, with each house requiring on average four panel deliveries. Other than the straw panel walls, the houses are very typical New Zealand construction, meaning that there is no specific training required for their installation, and the majority of the fixings are timber to timber, with no specific tools required.

The earth plaster system used for the wall panels is more airtight than regular construction, and the straw-bales provide improved insulation when compared to the building code, meaning that the homes are significantly higher performance than standard New Zealand construction.

In summary, Modcell® provides a great example of how structural engineers can intervene higher up in the supply chain to create products that help shift the construction industry towards more regenerative practice. This isn't about a single project, or even about a single product, it is about a group of people coming together, finding creative solutions to multifaceted problems and enabling beautiful, sustainable buildings to occur.

CASE STUDY 23

-

Base Bahay/Hilti Foundation, The Philippines

Our third case study looks into how material supply chain design at scale can yield multiple, layered economic and social benefits. All this while using an approach that harnesses the potential of abundant, renewable resources, demonstrating that construction practices can deliver substantial benefits across multiple dimensions. It shows what is possible when we design at operational systems level rather than project level.

The Philippines is a country grappling with a significant housing shortage. An estimated 6.5 million homes are needed to adequately house its residents, a task complicated by constraints such as financing and land availability. The challenges don't end there. The Philippines is also regularly hit by destructive disasters, which can wipe out up to 100,000 homes annually.

The challenge is therefore to create sustainable housing, at scale, which is durable and disaster resilient. Base Bahay was initiated by the Hilti Foundation to provide a holistic approach in addressing the global housing gap while helping the value chain and providing a positive impact on the environment. This innovative solution aims to build comfortable, affordable, disaster-resilient, and ecologically friendly homes which positively impact the community.

Project details

BASE works with the supply chain to create homes constructed from bamboo. The houses are made through the Cement-Bamboo Frame Technology that utilises prefabricated bamboo stud panels and can be quickly constructed (Figure 5.14).

Figure 5.14: Bamboo stud frame house being constructed

The bamboo is not taken from plantations but is instead harvested from local farms, where it occurs on the periphery, much like coppicing. In harvesting the bamboo, it leaves the plant to regrow on a 3-year cycle. By utilising bamboo, it enables farmers to generate a new income stream, while not negatively impacting available farmland. This material is therefore an abundant, renewable resource.

The treated bamboo is prefabricated into bamboo stud panels (using timber to create the top and bottom rails) (Figure 5.15). The duplex houses typically take five weeks to construct:

- Two weeks to dig the foundations
- (in parallel with digging the foundations) Two weeks to make the panels

- One day to erect the walls and floor
- Three days to construct the roof
- One to two weeks to apply finishes

Figure 5.15: Construction of bamboo duplex

The engineering side of BASE is led by Luis Felipe Lopez, who worked with bamboo as a structural engineer extensively in Colombia, and then relocated to the Philippines following a massive Typhoon in 2013. The first phase of his work was to create affordable and scalable treatment methods that worked in the Philippine context. By 2017, he had succeeded in this goal, laying the groundwork for the next phase of work: constructing homes.

With an ambitious target of constructing 10,000 homes by 2025, the initial progression was promising. The number of homes constructed each year escalated from 25 in 2014, to 100 in 2015, and 350 in 2019. In 2020 they were aiming for 500 when the global pandemic hit, significantly reducing the number they could build and the momentum of the project. Last year (2022) they were back up to 200 houses, not at the level they were pre pandemic, but they are looking to keep scaling the production in a sustainable manner.

Aiming higher in the system

One of the biggest problems with bamboo is insect infestations. BASE has worked with local farmers and

Figure 5.16: Treatment of bamboo culms

partner facilities to create an affordable and reliable process to significantly reduce the risk of beetle and termite infestation, making the houses much more durable (Figure 5.16). The expertise developed in this partnership is now shared with other treatment works, enabling the scaling of the use of bamboo, but the original treatment works, which are run as a collective by the farmers, are funded on a reciprocal agreement, where the farmers pay for the expertise of BASE in bamboo, which are then used in the community building projects. This symbiotic arrangement fosters community development and reinforces the use of local, abundant, renewable resources, providing a blueprint for other communities to follow.

The transition from harvesting and selling untreated bamboo (which holds little value) to offering prefabricated, treated bamboo walls represents a significant upscaling in the value chain. Farmers are able to sell a higher-value product, which enables them to invest and develop their

Figure 5.17: Testing on bamboo sections

supply chains. At the same time, working with BASE helps to ensure that the bamboo harvesting process is mutually symbiotic. Improving the quality of their product increases farmers' income, helps transition towards a supply chain that is enhancing the places it touches. This approach shows how we can intervene higher in the system in order to create cascading benefits to society and ecology.

Luis has worked collaboratively with local and international institutions, exchanging access to testing facilities (which universities have but are expensive to run) for publications, which are one of the main outputs for universities. As a result, not only is research able to occur, but the knowledge is shared, creating a win-win through collaborations.

BASE also builds special structures that provide economic empowerment and help the community to be self-sufficient. Apart from offices and community centres, BASE has built a Weaving Centre in Batangas (Figure 5.18), and a Multipurpose Building that will serve as a post-harvest

Figure 5.18: Kanya Kawayan Weaving Centre, Batangas

facility in Bagong Silangan, Quezon City. BASE is currently working with like-minded organizations to build a market village, school, and food processing centre.

Beyond the project, Luis and the BASE team have been sharing their learning and helping other countries adopt similar strategies. They are working with Habitat Nepal, providing technical assistance on the design of similar structures for wind and earthquakes. They are also working with Fiji to create houses that are adaptive to climate change.

Moreover, BASE has been working with the Association of Structural Engineers of the Philippines on the inclusion of bamboo to the National Building Code, which is also the objective for Nepal and Fiji. In doing so they are seeding change in how the construction industry operates in several countries.

CASE STUDY 24

-

Lee Sixth Form Centre at Ralph Allen School, UK

Our fourth case study is a little different. It highlights how the collaborations we form on a single project can create profound impacts that ripple across the industry. The story begins with the design of Lee Sixth Form Centre at Ralph Allen School in Bath, where James took the project from the bid to completion of Stage 2 (Figure 5.19). Here's how James tells the story:

Over the 15 years that I have worked in industry, I have decided, following significant reflection, that of all the projects I've worked on the Lee Sixth Form Centre is my favourite. The project is a new two-storey building constructed on an existing Redgra (a form of aggregate-based playing field for sports like hockey), which was past the end of life and no longer safe for use. The project is made from CLT and has won numerous awards, including a National RIBA award and Civic Trust award in 2014. The project architect was Feilden Fowles and the structural engineer was Ramboll.

My involvement in the project was from bid to completion of Stage 2. Due to other commitments, I was sadly unable to see the project through to completion. Despite this, of the 30+ projects I have worked on it remains my favourite. It is not the biggest. It is not the most prestigious. It did not win the most awards. It is not, arguably, the most sustainable. But it is my favourite.

If we were only to consider the project in isolation, while it is a beautiful, well-proportioned, well-loved building, it

Figure 5.19: Lee Sixth Form Centre at the Ralph Allen School, Bath

would not be my favourite. It has become my favourite as I have reflected on my relationship with it, which as you will see, stretches far beyond the project itself.

Sourcing local materials

The building has shallow foundations that sit directly on limestone bedrock. In the early stages of the project, when we carried out the site investigation, we dug a number of trial pits and decided to carry out testing. From the trial pits we extracted fractured limestone of varying sizes (Figure 5.20). Rather than see the limestone as a waste material, we explored different uses for it. Finally settling on looking to create a polished lime concrete floor slab using the limestone as aggregate which would make the locally sourced material legible in the final design (Figure 5.21). We carried out significant testing, making a variety of samples. Sadly, following testing and subsequent conversations with the client, the outcome was perceived as too risky, and the floor slab was cast from concrete[5.8].

Simultaneously we extracted soil to see if it could be used for a rammed earth wall. Initially the proposal was to use the earth to construct the internal spine wall which

Figure 5.20: Extraction of limestone from trial pits

Figure 5.21: Polished lime concrete floor slab sample panels

Figure 5.22: Testing the rammed earth — and the finished rammed earth waill

supported the first floor and roof. However, following testing, it was agreed that the wall would be constructed, but as a non-loadbearing dividing wall (Figure 5.22). Furthermore, due to the nature of the site-won material, the wall required significant cement stabilisation, reducing its sustainability credentials.

At one level, both of these explorations can be seen as failures. We expended time and effort in developing solutions which were not utilised. And the final construction, while made from bio-based materials (CLT and glulam) had a much bigger impact than we had hoped for.

But as you will see the cultural ripples from this project were extensive.

Long-term relationship

There is some very important context to this project that I have held back from you so far. Designing the Lee Sixth Form Centre was not, in fact, my first connection with the school.

Back in 1989, I joined the school in year 7, and attended all the way to the end of year 13 when I left with my A-Levels, ready to head off to university.

In 2001 I returned to the school. I had entered the Henry Palmer Award, an ICE initiative to take engineering into schools, a prize for which I was commended. Since then, I continue on an ad-hoc basis to work with the school. I have attended careers fairs, given talks to students, and run mock university interviews.

But my story doesn't stop there.

In 2015 the first of my children joined the school.

So, as you can see my relationship with the school, which is now more than 30 years old, continues in different forms. And the Lee Sixth Form Centre was just one of many meaningful interactions which we have both found beneficial (although I suspect I have gained more than they have).

When I was selected to bid for the project back in 2009 (the first time I had been asked to do so) my directors clearly knew what they were doing, sending an alumnus! But I hope they also saw the value in the relationship.

Culture

Working on Ralph Allen School had a profound impact on me (or perhaps more specifically, working with Ellen Grist, an Engineering Doctorate Researcher who was based at Ramboll, and Fergus Feilden, the architect on the project, has had a profound impact on me).

Ellen joined Ramboll in 2009 and was there for four years looking into how we take novel materials and enable them to become utilised in industry. Her work was wide-reaching. She carried out an in-depth study of lime concrete. Created a number of design guides on materials from reused steel to straw-bale. And she approached the problems using systems thinking, something I was new to at the time.

Without Ellen's work I would not have taught a unit on sustainable construction. We would not have been able to carry out the research we did on the Lee Sixth Form Centre. And I doubt I would be where I am today. The impact of her work on the trajectory of my life cannot be underestimated.

Fergus Feilden was at the time growing his very young architecture practice, Feilden Fowles. This was the only project we worked on together, but he has contributed in a number of different ways. He has taught on a number of my units. His practice studio (which is demountable and made from timber) is a case study in one of my timber text books[5.9]. The work we did on the Lee Sixth Form Centre has impacted on his design practice, which has continued to rightly win multiple prestigious awards.

But the cultural impact does not stop there. For the school, we (Ellen and I) produced an exhibition on different sustainable materials, which was then displayed at the Architecture Centre in Bristol (Figure 5.23) and then again at Bristol's 'Create' environment centre. The work we did on earth design, over time, morphed into my teaching notes, which are designed to enable practising engineers to design from earth, and I know are shared around engineering offices by my graduates. And in a moment of things coming full cycle, one of my graduates, who now works for Ramboll, recently asked me to send them a copy of the notes as they are creating an inhouse design guide on rammed earth for the company.

So, while the use of rammed earth (and lime concrete) on that project did not end up going to plan, it is amazing to see the ripples of that project as they spread out through the cultures we create.

Figure 5.23: The materials exhibition at the Architecture Centre, Bristol

Aiming higher in the system

To understand the Lee Sixth Form Centre in the context of local materials, long term relationships and cultures, is to understand it at the design level. But it is also useful to acknowledge how this project, and my work with Ellen, helped me see how I needed to aim higher in the system; at the operations level.

It was never enough for me to design sustainable projects (as I conceived them at the time). I wanted to change the industry in which I worked. This change started when I developed a new unit for Bristol University, called Sustainable Construction, which I first conceived back in 2007. However, it didn't come to life until 2011 following the work Ellen and I did together, which, alongside some other projects, was embedded in the Lee Sixth Form Centre project. Lecturing on the unit enabled me to teach hundreds of students about non-conventional materials, sending them out into industry with detailed notes on earth and straw-bale construction. It led to me changing job so that I could learn more deeply about earth, straw and reuse. And then changing job again to focus more fully on teaching and impacting industry. It led to the publication of books on timber design[5.9, 5.10], papers on lime concrete[5.11] and a published case study on how we failed to use innovative materials[5.8], and of course, the publication of this book, which I hope will further shift the conversation on sustainable and regenerative design.

Your story doesn't have to be the same. In fact, it will be different. But don't underestimate the impact thinking regeneratively about projects can have. I have found my way by feeling in the dark, we hope that this book will help others to see more clearly.

Summary

James' involvement with the Lee Sixth Form Centre at Ralph Allen School is an example of how even a single project's collaborations and interactions can have a far-reaching impact across various operational contexts within the industry. Ideas that formed within the design team extended far beyond the confines of this individual project, influencing industry practices on a broader scale. The professional relationships developed throughout the project endured beyond its completion, demonstrating the potential longevity of collaborative partnerships.

His experiences working with locally abundant and renewable resources were the seed for teaching that has impacted many students, and now this teaching reaches further — through the books James has published — which in turn is building our industry's capacity to turn away from extractive material use. We also see from James working on the school he and his children went to, a glimpse of what it could be like for engineers to be more involved in the communities they live in.

5.3 INTERVENING HIGHER IN THE SYSTEM

The four case studies in this chapter show different ways that structural engineers have gone beyond project level to influence how their part of the wider construction industry operates. To use the language of this book "they have intervened higher in the system to change the construction industry at operations level".

To draw together the ideas in these case studies and to reinforce concepts from this book so far, we have assembled a set of questions, grouped together according to different types of operations-level changes we could be making (Table 5.1). You can think of this as a brief guide for structural engineers on how to intervene higher in the system to transition to a more regenerative industry.

Table 5.1: Questions we could be asking at the operations level

Operations-level change	Questions to ask
Capacity to self-organise	• How can we create a working culture in our industry that continuously learns from ecosystems and communities? • In what ways can ecosystems and communities be brought into the design process? • How can we promote interdisciplinary working to bring a wide range of expertise into the design process? • What channels can we use to share innovation, success and failures more widely?
Rules that govern system behaviour	• How can we create the conditions that allow for more experimentation and community-based design? • How can we create and distribute design guidance for new and rediscovered materials? • How can contracts, design codes and codes of conduct be updated to prioritise human and living systems' ability to survive, thrive and co-evolve?

Table 5.1: *continued*

Operations-level change	Questions to ask
Information flows in the system	• Just as a new stream of information on embodied carbon is influencing design decisions, how can we feed in information about ecosystem and community health? • How is information brought to life — not just numbers about species, but a more deeply understood and felt appreciation of the information? • How can new and rediscovered ways of working be more widely shared to influence designers and other decision-makers?
Feedback loops that balance and reinforce system behaviour	• How can real-time insights into the vitality of ecosystems and communities be integrated into design decisions, ensuring steady enhancement of existing conditions? • Can we simplify supply chains to create a clearer signal between ecosystem, community and designer? • How can the balancing feedback loops that limit destructive practice be amplified so that we pay more attention to them? • How can the reinforcing feedback loops that encourage regenerative practice be amplified so that we also pay more attention to those? • What design strategies can nurture deeper symbiotic relationships between communities and their ecosystems?
Stocks and flows of materials	• As we assess the health of ecosystems — from water quality to species diversity — and communities, from wellbeing metrics to trust indices, how can design solutions amplify these factors? • How can the system of material harvest (rather than material extraction) be brought into connection with the designer and the community that benefits from it? • What methods can grow and unlock the renewing and abundant materials in a system? • How can the wealth generated from construction be reinvested in the communities from which it is generated rather than being siphoned away?

5.4 CONCLUSIONS – AIMING HIGHER IN THE SYSTEM

In this concluding section of our chapter, we circle back to our initial question: How can a structural engineer facilitate systemic change at the operations level in the construction industry?

Recall from Chapter 2 that we can think of the construction industry as a hierarchical system. To make significant shifts within this system, we must intervene higher up in this system or, as we've said by analogy, higher up in the Systems Bookcase. In Chapter 3, we proposed that the role of the regenerative structural engineer is to facilitate the transition towards a regenerative construction industry, creating stepping stones for this transition. In Chapter 4, we identified how design could unlock this transformation, acknowledging that systemic operating rules and regulations could limit our impact.

In this chapter, we've risen up the Systems Bookcase to explore various operational contexts within our sphere of influence. These contexts encompass collaborations within the industry, supply chain dynamics, the civic sphere, and wider industry implications.

We've looked at four case studies, each intersecting these industry contexts, highlighting how structural engineers can leverage their power at the operations level to unlock the industry's potential for a more regenerative practice.

In every case study, the impactful change emerges from thinking beyond the scope of a single project — to thinking higher up in the system. Sometimes, systemic-level thinking was the objective from the start. In other cases, it emerged as projects progressed. Regardless of the origin, the strength of this approach can be seen when we operate at this level; connecting communities, ecology, and the spaces we impact with our work. We begin to shift our industry towards more significant use of abundant and renewable resources — revealing their potential by overcoming systemic obstacles — and we harness the power of what is trying to emerge from the situation on the ground, enabling the self-organising capacity of systems by removing barriers, and building skills.

By thinking on a systems level, regenerative structural engineers can play a significant role in transitioning us towards a more regenerative construction industry, steering us away from historically destructive practices. Yet, what ultimately governs the ambition for what we can achieve is the collective mindset that we hold in industry. This is the next level up in our Systems Bookcase, and it will form the focus of Chapter 6.

REFERENCES

5.1 Terzano K. *Human Dimensions of Civil Engineering: Context and Decision-Making for a Sustainable Future.* Boca Raton: CRC Press, 2023.

5.2 MASS Design Group. *Adobe Block (Rukarakara) Standards.* Available at: https://massdesigngroup.org/work/research/adobe-block-rukarakara-standards [Accessed: December 2023].

5.3 Modcell® Straw Technology. *Helping you build a more sustainable future.* Available at: www.modcell.com/ [Accessed: December 2023].

5.4 Pelly R and Mander T. 'The Nucleus at Hayesfield Girls' School, Bath: achieving low embodied carbon using renewable materials'. *The Structural Engineer.* 2014:92(10). Available at: www.istructe.org/journal/volumes/volume-92-(2014)/issue-10/the-nucleus-at-haysfield-girls-school,-bath-achiev/ [Accessed: December 2023].

5.5 Chatterton C. *Unlocking Sustainable Cities.* London: Pluto Press, 2018.

5.6 The Institution of Structural Engineers. *Design for zero.* London: IStructE Ltd, 2021.

5.7 YouTube. IABSE British Group. *Milne Medal 2021 Lecture "Less is more" by Margaret Cooke.* Available at: www.youtube.com/watch?v=Z9CpYUFbtlE [Accessed: December 2023].

5.8 Grist E et al. 'Innovative solutions please, as long as they have been proved elsewhere: The case of a polished lime-pozzolan concrete floor'. *Case Studies in Construction Materials.* 2014:(1). Available at: https://doi.org/10.1016/j.cscm.2014.01.001 [Accessed: December 2023].

5.9 Norman J and Thomson A. *Designing timber structures: An introduction.* High Wycombe: BM TRADA, 2020.

5.10 Norman J. *Structural timber elements: a pre-scheme design guide* (2nd edition). High Wycombe: BM TRADA, 2017.

5.11 Grist E et al. 'Structural and durability properties of hydraulic lime–pozzolan concretes'. *Cement and Concrete Composites.* 2015:(62). Available at: https://doi.org/10.1016/j.cemconcomp.2015.07.001 [Accessed: December 2023].

CHAPTER 6

How to think regeneratively: creating transition through mindsets

How do we create the mindset for a transition to a more regenerative built environment industry?

In Chapters 4 and 5 we explored the potential for project design and systems operations to create a transition towards a regenerative built environment industry. We discovered that while challenging how we design and challenging system-wide rules can bring about change in our industry, these approaches are ultimately limited by the mindsets that permeate the industry.

In this chapter, we go into a higher level of transformation, the realm of mindsets — the next shelf up on the Systems Bookcase introduced in Chapter 2 (Figure 6.1). Specifically, our aim is to propose mindsets that could catalyse a transition towards a more regenerative built environment. Then, having identified what these mindsets are, we explore what structural engineers can do to grow these mindsets in industry. So, how can structural engineers think regeneratively? And how can we embed that thinking into the profession?

Figure 6.1: The mindsets 'shelf' in the system

Regenerative design, as we established in Chapter 3, is an example of third-horizon thinking. From the context of our present horizon, envisioning this regenerative future can be challenging, so we return to our Living Systems Blueprint, also introduced in Chapter 3, for guidance.

Drawing from this thinking we propose three mindsets: interdependence, abundance and emergence. Together, these three ways of perceiving the world and the opportunities that surround us have the potential to pave the way for a built environment sector that not only ensures survival, but also enables the thriving and co-evolution of human and living systems.

To illustrate this we examine a case study that shows the transformative power of changing mindsets on our industry.

In the last part of the chapter, we introduce a toolkit designed to encourage these mindset shifts based on cultural transformation; providing a pathway for mindset change — a critical step in the transition to a more regenerative built environment.

6.1 MINDSETS FOR REGENERATIVE DESIGN

In chapter 3 we identified three characteristics of living systems that can help us understand how these systems survive, thrive and co-evolve. These characteristics can guide us in envisaging a regenerative built environment and can provide insight into how we might need to structure our industry to achieve this goal.

The three living systems' characteristics are:

1. Their high levels of interconnectivity
2. Their use of symbiotic loops to create increasing abundance
3. Their capacity to self-organise in response to environmental changes.

Drawing from these characteristics, we can imagine a built environment industry:

- Acutely conscious of its interconnections within all aspects of the human-living system, acknowledging these connections as the basis of resilience and survival
- Meeting its needs through integrated and renewing cycles of supply
- Actively unlocking the potential of each place we work in, allowing the elements there to evolve organically in response to the environment rather than rigidly controlling what happens

We called this vision our Living Systems Blueprint for Regenerative Design (Section 3.2.4).

In this section, we'll explore three mindsets that could unlock a transition to this future (Figure 6.2). We'll then go on to talk about the role that the regenerative structural engineer can play in building these mindsets.

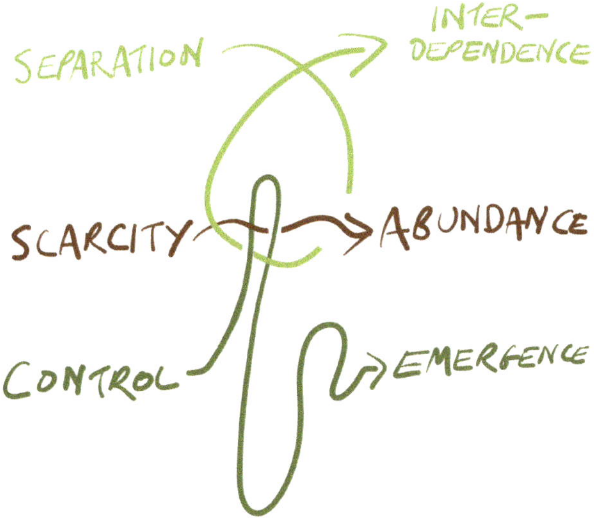

Figure 6.2: Three shifts in mindset to enable a transition to a more regenerative built environment industry

6.1.1 From separation to interdependence

The first mindset we'll explore is that of interdependence. This perspective enables the shift towards an industry that is more deeply aware of its interconnections with all facets of the human and living system, understanding these links as fundamental for resilience and survival.

If we look at how society in general, and specifically engineers in the Global North, interact with the living world, we can see that the prevailing mindset is of separation. This is not a new phenomenon. As cognitive anthropologist Jeremy Lent argues, this sense of separation began in societies which shifted from hunting to settled agriculture millennia ago. From then on, those societies didn't move with and follow the flow of the herd: they had to stay in one place and hoped the living world would give them what they wanted. According to Lent, the living world began to become a separate thing, one to be feared and distrusted[6.1].

While the phenomenon of separation from the living world began long ago, we can see how far it's gone by just how 'non-living' (i.e., devoid of life) many of the built environment spaces we create are. From car parks to shopping centres to many offices and homes, the built environment is full of places where there is very little sign of life (Figure 6.3).

Our point here is not to suggest we return to hunting and gathering! Instead, we need to acknowledge that given how far away we have pushed the living world, it is not surprising that we have a sense of separation. From this perspective we can ignore the complex web of relationships that tie us and

Figure 6.3: A city centre almost devoid of visible living things

the rest of the living world together, and in doing so, make decisions that may benefit us but at the cost of living systems. This mindset of separation is arguably one of the root causes of our ecological emergency[6.2].

But the mindset of interdependence isn't just about recognising connections, but about understanding and feeling that the welfare of one element is intrinsically tied to the welfare of all others. It's the recognition that if parts of the living system are suffering, we too will suffer — a reality made increasingly evident through the climate and biodiversity crises. This mindset also acknowledges that if human systems are in distress, there are consequences for our individual and collective wellbeing. This echoes the findings presented in the book *The Spirit Level*, which suggests that societies with higher levels of inequality experience lower overall happiness, even amongst the wealthy[6.3].

But as we have seen from the Systems Bookcase model, mindsets are powerful. If the construction industry could collectively adopt the mindset of interdependence, it has the potential to cascade down through the rest of the system. It would shape the rules, regulations, information flows and feedback loops at operations level. And these in turn would change what we can do in design.

For example, a collective mindset of interdependence could:

- Prompt individuals to broaden their perspective on the 'second site' as introduced in Chapter 4, examining more fully the impact of their work throughout the entire supply chain
- Catalyse the creation of new information flows concerning the positive and negative impacts of decisions, fostering greater transparency and accountability

- Lead to the development of rules and regulations that govern not only embodied carbon but also 'embodied biodiversity' and 'embodied social justice', integrating these important considerations into standard practice
- Establish a stronger imperative to design for the wellbeing of wider human and living systems when creating buildings, shifting the industry's core objectives

We are using the term 'embodied social justice' to mean the potential positive or negative impact that a supply chain can have on social justice issues — similar to 'embodied carbon', rather than the practice of seeking to understand social justice through a practice of embodiment. For the latter, refer to Rae Johnson's *Embodied Social Justice*[6.4].

In Section 6.2, we'll explore how we might inspire this change of mindset from separation to interdependence, but first, let's delve into two further mindset shifts.

6.1.2 From scarcity to abundance

The second mindset shift we look at is transitioning from scarcity to abundance. In our competition-orientated economy there is an ingrained belief in scarcity[6.2] — that there is not enough for what we need and that we must compete to get what we need. This scarcity mindset can manifest itself in different ways: in taking more than we need in case we run out; in fearing that other parties in an agreement will act competitively rather than collaboratively; in allowing fear of failure to impede the vision of what might be possible. This isn't to say that all competition is bad, but to consider what happens to our view of other people and resources when competition becomes our dominant world view.

In contrast, consider an abundance mindset, which can be inspired by a living system's capacity to create abundant webs of resources.

Figure 6.4: Example of an abundance mindset — a community food- and plant-sharing stall. People take what they need and replenish when they have spare items

For more on the role of interconnection in regulating human's use of common resources, see the entry on 'Commons' in Fleming's *Dictionary of Lean Logic*[6.5].

These riches are created from a network of organisms exchanging information and nutrients, and creating the conditions for life. This exchange is enabled and regulated through the 'interconnection' we frequently refer to in this book (Figure 6.4).

The shift to an abundance mindset means seeing the potential for human and living systems thriving. It means seeing the potential in what the living world produces to provide for our needs, and the potential for humans to contribute to that abundance. It is a mindset that recognises abundance comes in cycles, and that we need to work with the rhythm of these cycles. And it is a mindset that enables us to see the potential for thriving where at the moment there is only a desert.

Furthermore, an abundance mindset is not just limited to physical resources but extends to how we see our relationships with the people we work with and for. An abundance mindset enables trust, collaboration and, dare we say it in a textbook on structural engineering, love. Because to love you have to overcome the fear of loss, but when we love something, it can thrive.

In this way we can see that an abundance mindset actually builds on the mindset of interdependence.

As with the interdependence mindset, an abundance mindset could cascade down through the rest of the construction industry, creating the potential we see in the people and resources we work with, shaping the rules for how we maintain balance in these systems and ultimately influencing the countless decisions we make on a design project.

For example, a collective mindset of abundance could:

- Nurture a culture of collaboration rather than competition. Which in turn, could:
- Allow engineers to share information more freely with each other about how to use an abundantly available natural material. Which in turn, could:
- Enable new breakthroughs to be made in how resources can be used symbiotically. Which in turn, could:
- Enable the use of materials on a project that actually enhance, rather than deplete the site of origin

Just like the shift from separation to interdependence, the transition from scarcity to abundance has the potential to transcend the entire industry, influencing its operations, regulations, and the resulting built environment.

So far, we have proposed two mindset shifts: from separation to interdependence; and from scarcity to abundance. Let's delve into our third mindset shift.

6.1.3 From control to emergence

The third mindset we will consider is moving from control to emergence. Modern practices within the construction industry are often associated with seeking ever greater levels of certainty. Each detail of a project is highly planned and coordinated. Each construction process has a tightly controlled method statement. The aim of course is to minimise risk and increase productivity. As engineers we like certainty — in fact, we are often paid to provide certainty (and we take out insurance in case we get that wrong).

But the human and living systems we want to help survive, thrive and co-evolve, are complex systems. This insistence on control is fundamentally at odds with how we understand complex systems to operate. As we learned from our exploration of living systems, we can't fully predict what complex systems are going to do, and so we can't fully control them. Instead, we must work with them, adjusting and adapting our designs and operations based on their responses and shifts over time. This understanding underpins the Continuous, Place-based Design Process we defined in Chapter 4.

In contrast, an emergence mindset aims to harness and work with, rather than against, the way living systems operate: working in a self-organising and adaptive manner (Figure 6.5). Living systems evolve and adapt in response to changing environmental conditions. They allow for a degree of variance and uncertainty, and through interconnection and feedback loops, evolve to create better, more resilient responses to the specific ecological conditions of a place. This emergent behaviour is not top-down imposed but rather arises from the interaction of the system's parts through many layers of interconnection and feedback. This emergent form of design has led to the evolution of the incredibly complex, dynamic and balanced living systems we see around us today.

Figure 6.5: Big Jarrah (named after the railway sleepers used to build the timber frame) by Yes Make. The incorporation of reclaimed materials, such as these windows which were surplus to requirements on another project, requires an emergent approach that responds to opportunities as and when they come up

Embracing an emergence mindset in construction is not about ignoring risks. Rather, it is about learning how to work with a greater degree of uncertainty and unpredictability in our approach to work, unlocking the potential for design solutions to emerge that are a better fit to the places and the systems we are in.

It is about noticing when we are trying to impose certainty onto a situation and asking instead, what would happen if we created the conditions for the appropriate solution to emerge? It is about shifting from deterministic methods to ways of working that are more iterative and responsive to the complex signals we are working with. The emerging ideas may not align with our initial conception, but they are likely to be much more closely attuned to the system's requirements. The approach may not draw upon our extensive experience from past projects, and mistakes may indeed be made, but the learning derived from these mistakes is more likely to be absorbed by the people involved.

In this way of working, the role of engineers becomes more of a delicate balancing act: enabling experimentation and innovation on one hand, while exercising appropriate intervention on the other.

The shift from control to emergence, like the shifts from separation to interdependence and from scarcity to abundance, has the potential to transform the entire construction industry.

For example, with an emergent mindset we might set much more flexible goals for industry initiatives and projects, enabling more specific aims and objectives to be developed in response to the needs of the place, stakeholders and ecosystem. Doing so steers us away from being too prescriptive and opens up the possibility of more appropriate innovative solutions emerging. It steers us away from one-size-fits-all-but-doesn't-really solutions, to initiatives that are more tailored to local operating circumstances.

Or, an emergence mindset could encourage the development of codes of practice and building regulations that can be more flexible and adaptable in how they allow things to change. This in turn enables design in projects to better adapt to the conditions of the place.

Similarly, an emergence mindset could encourage a much greater flow of information. In an emergent system, shared information is critical to ensuring all parts of the system can work with what is emerging. It encourages collective sharing of knowledge, greater transparency and therefore better decision making. This approach might lead to more collaborative approaches that involve more real-time changes.

6.1.4 Conclusion: interdependence, abundance and emergence

In this section we have looked at how mindset shifts can unlock a transition to a more regenerative built environment industry. Returning to our Systems Bookcase model, if we put interdependence, abundance and emergence on the mindsets shelf of the middle bookcase, these mindsets have the power to unlock and guide everything that happens at the level of operations and design (Figure 6.6).

Let's be clear, it is not that the mindset shift on its own makes the change; rather, it provokes new perspectives. When we start to see the world in a different light, we start questioning the world around us: Why do we do things in this way? Is there a better way? It allows us to imagine a different possibility.

The role of the regenerative structural engineer is to enable the transition towards a regenerative built environment. We do this by noticing what is stopping that change, and what

Figure 6.6: Mindsets for transition to a regenerative construction industry, shown on the Systems Bookcase

we can do to enable that change. When our perspectives are framed by the prevailing lenses of separation, scarcity and control, it's hard to imagine a regenerative built environment industry. But if we can change the way we look at the world, we begin to notice how our system operates, and start to recognise the change we need to make.

All that said, changing mindsets isn't as simple as a thought exercise. It requires us to do something a bit more rigorous to stop us just reverting to our default mode of thinking. Here, culture plays a powerful role. In the next section we'll explore how to use culture to shift our collective mindsets in a regenerative direction.

6.2 EMBODIED CARBON

As we have highlighted throughout the book, regenerative design is about much more than just embodied carbon. Using the idea of planetary limits, engineers need to consider the impact they have across all the different boundaries. But, in this Horizon One space, how can we learn from the recent story of carbon emissions in our sector (and specifically embodied carbon) to see how we can then consider all the planetary boundaries in a holistic systems model as we look to Horizon Three?

The following case study is not an attempt to capture the full story of embodied carbon. Instead it is James' journey through calculating embodied carbon. It is based on his experience both in industry and in academia, and it is both incomplete and reliant on imperfect memory!

CASE STUDY 25

–

An embodied carbon journey

Back in 2000, when I started working in industry, I was aware of the need to reduce greenhouse gases, and we designed buildings with passive cooling and reduced operational carbon. At that point the (incorrect) assumption was that the embodied carbon was a tiny piece of the pie chart of carbon associated with buildings (Figure 6.7).

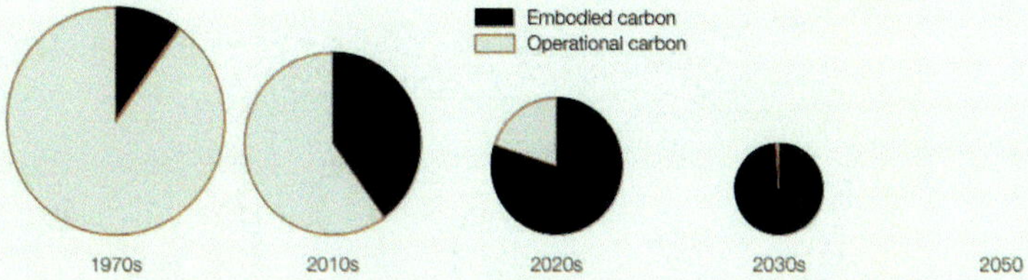

Figure 6.7: Changing importance of embodied carbon (based on approximated data)[6.6]

I have a vague memory, discussing and debating whether we should know the embodied carbon on the projects I worked on between 2000 and 2010.

I first became aware of what information was available about embodied carbon in 2009 when Ellen Grist joined Ramboll as a researcher under my supervision (Case study 24). She was calculating the embodied carbon and energy in different types of concrete[6.7] using the Inventory of Carbon and Embodied Energy (ICE)[6.8], produced at the time by the University of Bath. The ICE was a daunting amount of information!

Around the same time I organised a Research and Development conference and one of the presenters, Katie Symons, presented how she had calculated the embodied carbon of different schemes (Figure 6.8).

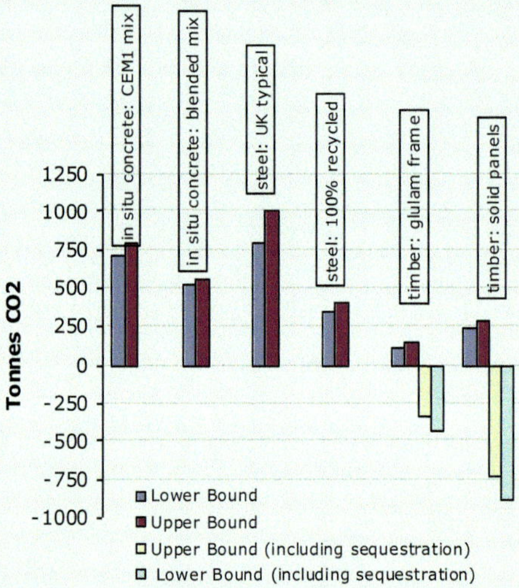

Figure 6.8: Assessment of the embodied carbon of different materials for a specific building

Based on this information, in 2012 I started teaching students at the University of Bristol how to calculate the embodied carbon on projects. However, while the theory was simple, calculating the quantity of materials was laborious.

When I joined Whitbybird (which later merged with Ramboll) most drawings were done in 2D; 3D drafting was just starting to take off. When I left Ramboll in 2012 a new approach to drafting and information was materialising; BIM. When a 3D model is combined with information (e.g., the materials that are being used) then you can quickly and efficiently calculate the volumes and masses of materials used on a project.

When I left industry in 2015, calculating the embodied carbon was beginning to become a reality that could be carried out not just by specialists and researchers but by anyone with a BIM model and access to the ICE database.

While I wouldn't be able to capture all the different ways things have progressed since 2015 there are a few standout moments for me personally.

Buro Happold produced a study of the carbon intensity of the different solutions for a typical office building, which highlighted how not just materials, but column spacing and foundations, have a huge impact on the carbon footprint. It also highlighted what a 'typical' carbon footprint might look like, and more excitingly for me, that it was possible to build with a much lower carbon footprint by making different decisions. They also shared this study via the IStructE for free (Figure 6.9)[6.9].

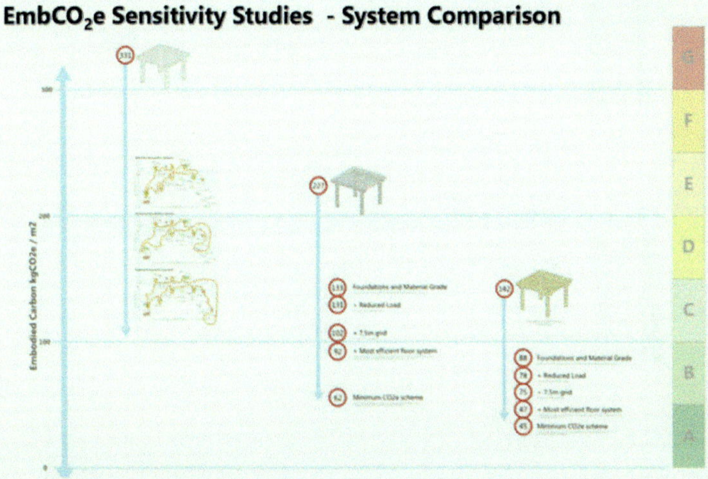

Figure 6.9: Extract from Buro Happold Embodied Carbon Study

In the same year, the IStructE published *How to calculate embodied carbon* which combined the ICE database and the RICS *Professional Statement on Whole Life Carbon Assessment for the Built Environment* (2017), simplifying the data and providing a clear methodology. This valuable information was made available as a free-to-access resource.

Around the same time, the IStructE turned their attention to the Climate Emergency, employing Will Arnold as the Institution's Head of Climate Action.

Since 2020 Smith and Wallwork have published an annual materials and carbon report[6.11].

Simon Smith from Smith and Wallwork came and spoke to my students. And one of the things he presented was not the carbon for a specific project, but the carbon specified by his practice. Virtually all of it. From 2014 (Figure 6.10). He was also able to show the mass of materials specified, and how this changed year on year. Suddenly he was able to present the carbon impact of his whole practice.

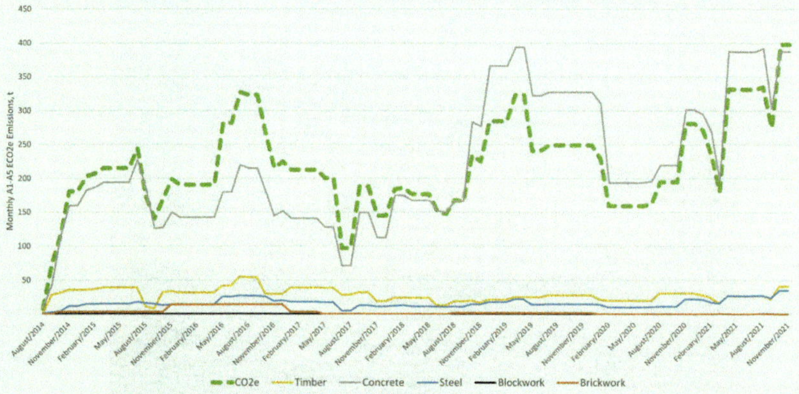

Figure 6.10: Monthly material usage and carbon emissions for Smith and Wallwork as a practice from 2014–2021

Elliott Wood also shared their Structural carbon tool[6.12], which enabled engineers to quickly estimate the embodied carbon in their building (Figure 6.11). This could be used at the conceptual design stage, so that even before the BIM model had been created, the carbon impact of different solutions could be assessed. Elliott Wood

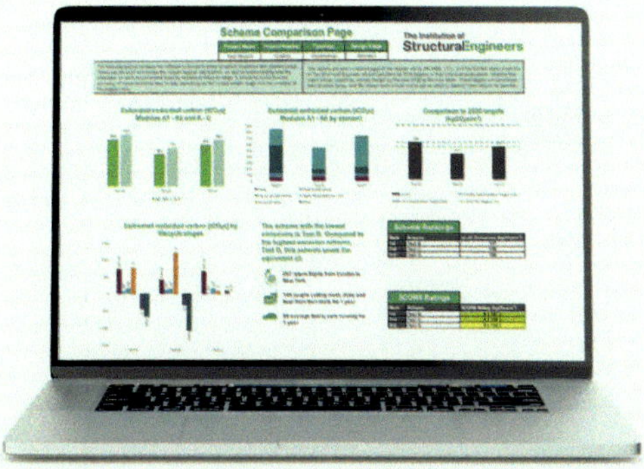

Figure 6.11: Structural carbon tool

decided to share this tool with all engineers, for free, when they could have kept it for themselves.

However, knowing the carbon footprint, while important as a starting point, is not enough — we need to reduce the carbon footprint. Elliott Wood and Buro Happold's contributions helped to start this conversation, but the IStructE's *Design for Zero*[6.6], pushed it forward providing clear and practical advice for engineers to reduce the carbon in their projects. Alongside this was the proposed SCORS rating, which is a way to rate projects (Figure 6.12)[6.13].

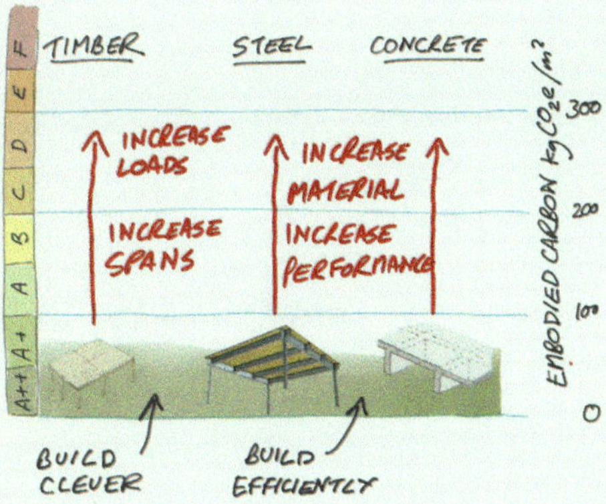

Figure 6.12: The SCORS rating system challenges us to design with lower carbon

And now, with the work of Will Arnold and the IStructE we have the emergence of rules, whether as Part Z of the building regulations (Figure 6.13), or in a different form. And AECOM have been appointed by the Department for Levelling Up, Housing and Communities to review the impacts to business of carrying out whole life carbon assessments.

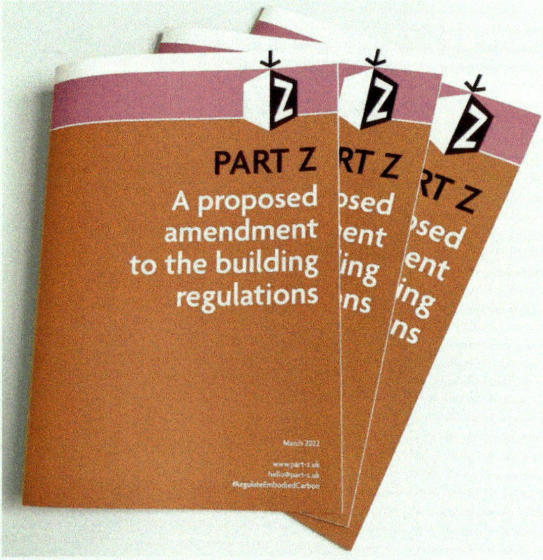

Figure 6.13: Part Z: An industry-proposed amendment to UK Building Regulations 2010

Mindsets

This case study demonstrates all three mindsets to greater or lesser degrees.

Interdependence: The story of global warming is one of separation and interdependence. It is a story of denying that humans can change the atmosphere and realising the catastrophic impact warming could have, how every living thing on the planet could be impacted and how, as a result we are all connected.

Abundance: The story of abundance here is different. In reflecting on the story of how we got to where we got to, there is a recurring theme: people either doing things for free or giving away their work for others to use. This captures an abundance mindset. It would be possible for engineers to see an ability to calculate embodied carbon quickly as a USP, a marketable commodity that they could use to maximise profit. The scarcity mindset. But in light of the climate emergency, time and time again, I have seen people share information, tools and their time, without remuneration (although they will have received other forms of capital, as by sharing their work they position themselves as leaders in the field).

Emergence: The final mindset that we see is emergence. There was no masterplan. No one sat down 15 years ago and said: "We need to create Part Z and here's how we'll get there". Instead, different organisations and people shared their information, and allowed others to use it as a creative catalyst for their own work. Without these grass roots initiatives, which by sharing information, created a collective information set that could be utilised, a top down approach almost certainly could never have happened. We are now approaching regulation of carbon, rules being put in place; at first to declare what the carbon footprint of projects is, but then with a view to reducing the carbon footprint — setting goals that we can then work towards.

We can also see these mindsets at work over and over again in the case studies we have presented throughout this book.

6.3 TRANSFORMING MINDSETS THROUGH CULTURE

In our experience, one of the most powerful tools available to us in our industry is the ability to affect cultural change. We see the impact of culture in the domain of health and safety and in programmes like the Get It Right Initiative[6.14], which seeks to create a culture of error-avoidance in construction. As Frederik Laloux puts it, in his book *Reinventing Organisations*[6.15]: "Culture is how things get done". It is the unwritten operating manual for an organisation, that permeates all actions.

Drawing on the ideas of Ken Wilbur, also cited by Laloux, we can see culture as the collective mindset of a group of people. Therefore, creating a collective mindset of interdependence, abundance and emergence equates to creating a culture that embodies these values.

Culture is, in effect, the collective mindset of a group of people (Wilbur). To try and cultivate a mindset of interdependence, abundance and emergence in our industry is to actually create a culture of these things.

The good news is that we get to influence culture every day. And because of the high position that mindsets take on the Systems Bookcase, changing culture can have a huge impact on the organisations we work in and with.

Admittedly, culture can be hard to pin down — it can feel nebulous. But there are well established tools for dealing with it. One that Oliver has worked with in many built environment contexts is the Cultural Web by Gerry Johnson and Kevan Scholes. In the next section we'll use this tool to think about how we can create a culture of interdependence, emergence and abundance in our industry.

6.3.1 The Cultural Web

The Cultural Web[6.16] is an effective tool for defining and thinking about how to shift an organisation's culture. By organisation, we can mean anything from a small team to a large company, to industry as a whole. The tool provides us with six different lenses through which to examine our current ways of working (Figure 6.14). The combined view through these six different lenses provides a more tangible picture of the organisation's culture.

In this section we will focus on culture at the scale of companies, as this is the realm over which most readers are likely to have some influence, but the approach we take here can be applied at a wider scale too.

By examining how we work through each of these lenses, we can see how our collective behaviour supports or undermines the cultivation of a mindset of interdependence, abundance and emergence.

Let's look at each of these lenses in turn.

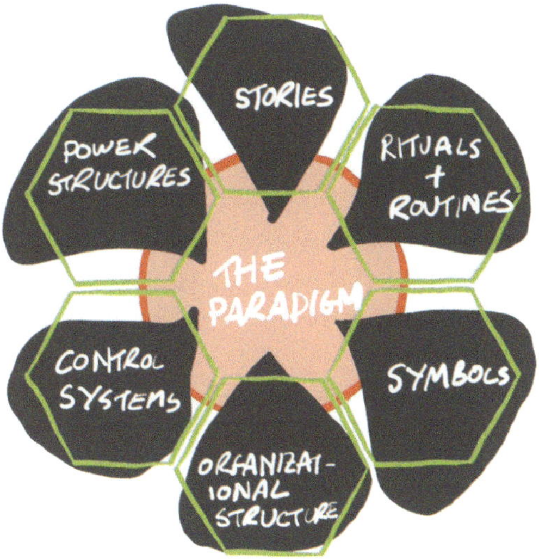

Figure 6.14: The Cultural Web

6.3.1.1 Control systems

The first lens in the Cultural Web we'll consider is that of 'control systems' — in other words what we decide to measure and manage. In structural engineering, these commonly include timesheets, project budgets, programmes, and demarcations such as a red line drawn around a site boundary, symbolising the bounds of design process control. Intellectual property controls who gets access to what information, while contracts essentially shape how we behave toward one another.

Of all the lenses in the Cultural Web model, we see 'control systems' as offering the greatest opportunity to nurture mindsets of interdependence, abundance, and emergence.

The red line

We create a line of control when we draw a red line around the site boundary and say what is inside concerns us and what's outside doesn't. The interdependence mindset challenges the isolation of the construction site from other places we impact in the design process.

The push to decarbonise our industry has introduced a new metric, embodied carbon, which expands our zone of control to include more of the material acquisition process. Circular design[6.17] further extends the boundary, leading us to consider the post-use lifecycle of material. By consciously broadening our design focus, we extend the project boundary to encompass more of the places influenced by construction. This shift changes what we pay attention to, and consequently, what we care about.

From control to care

An abundance mindset is about recognising the potential of the people and systems around us. It requires us to shift from a scarcity mindset, often stemming from a need to control, to recognising the potential in collaboration, sharing and creating independence.

But what if, instead of trying to tightly control projects or people, we framed our work as 'care'. This can be compared to raising children; we aim to support and guide them so they can grow independently and flourish on their own. In a design context, this might look like giving team members the autonomy to explore and innovate, trusting that their creativity will enrich the project.

Knowledge

We can also seek to create an abundance mindset in our approach to sharing knowledge and resources. With a scarcity mindset we might hoard information or resources for fear of losing advantage. With an abundance mindset we more freely share and collaborate, encouraging greater innovation, learning and adaptation — contributing to more resilient teams and connected networks.

A more agile approach

There is a precedent for adopting an emergent mindset in design: it is called 'agile project management'. In an agile approach we seek to take an experimental perspective, rather than trying to control all of the parameters. This allows us to work much more adaptively with the complexities of the system. For projects in the built environment, this becomes possible when we take a longer-term view of design — creating interventions over time that we modify in response to continuous testing.

These suggestions are not intended to be comprehensive but are intended to provoke your thinking about how we can change the way we exercise control to help nurture mindsets of interdependence, abundance and emergence.

6.3.1.2 Stories

Typically, whenever you join a new company or team, people start to tell you stories: tales of how projects went; individuals celebrated (or criticised); key milestones hit (or missed). In structural engineering we tend to like to tell stories about clever ways we've solved problems, work delivered under pressure and stories about how difficult a particular consultant or contractor has been.

The 'stories' lens of the Cultural Web acknowledges the powerful role that stories play in reinforcing the culture in our organisations. Stories give us clues about what we collectively desire and what we want to avoid. The stories that we choose to tell and the creation of opportunities to tell these stories are a powerful tool for shaping culture.

Therefore, one of the ways to cultivate mindsets of interdependence, abundance, and emergence, is to change the narrative of the stories we tell.

Stories of interdependence: These could focus on an exploration of the supply chain that helps to build interconnection. The stories could draw attention to symbiotic relationships that you are working with and potentially creating. It could include stories of inspiration coming from our community or nature. It could also include us talking about our connection with the living world.

Sometimes interdependence comes down to the language we use. Do you say you were late because you were stuck in traffic or because you *were* traffic? Do you say management is restructuring or we are restructuring? Do you say the project did harm/good or do you say we did harm/good? Putting yourself into the story helps to create a culture of interdependence.

Stories of abundance: These might draw attention to collaborations — what happens when we trust others, when we share information or resources. It could include recognising the distinct qualities of a local waste material, harnessing them, and sharing this information with our community, so others can use it as well. It could involve collaborating with a number of local practices to create an inventory of unused buildings, waste materials, and local skills and ways of working.

Stories of emergence: Rather than emphasising project completion, can we tell stories about an ongoing relationship with place, about being engaged with a community, with an ecosystem and showing what we are learning from this longer-term relationship? Can we draw the lines backwards from what we have done to what others did, celebrating and giving credit to those that have inspired us?

Stories of transition: One of the key ideas we keep reinforcing in this book is that the role of the regenerative structural engineer is to help create the transition to a regenerative built environment sector. So make the specific change you are seeking to make (the specific part of the industry you are trying to change) the key part of the story. Not just what you did, but why.

James actually shared the process of writing this book over a series of blog posts for exactly this reason[6.18], and Oliver regularly collects stories at the workshops he runs (Figure 6.15).

Figure 6.15: Oliver collecting stories at a Buro Happold workshop on regenerative design

By being conscious of the stories we tell, we start to change the collective narrative about what we hold important.

6.3.1.3 Rituals and routines

The next lens in the Cultural Web is that of rituals and routines. Through this lens we try to characterise the ways that people go about their work on a daily basis. These are the things you might expect to see people doing in the office, on site or in a team meeting.

Examples of routines and rituals we might recognise in structural engineering are: design team meetings, lunchtime training sessions, resourcing meetings, project reviews, requests for information from site, a regular night for social gatherings, celebration of laying the first foundation, or topping out.

These patterns of behaviour are signals of what is acceptable within the collective mindset of the organisation. But if we shift our patterns of behaviour we can shift mindsets. Here are some suggestions for shifts that might help us nudge mindsets in the direction we're talking about:

Routines that value biodiversity: In construction 'SafeStarts' are already commonplace, whereby a meeting involving say, more than three people, must begin with a moment to talk about health and safety. What if we developed similar routines to talk about the vitality of the living world in the places we are working? By repeatedly drawing attention to the vitality of the world around us, we start to value that vitality as a mindset.

Rituals that acknowledge land and place: In many countries it is becoming increasingly common, at the start of an event or presentation, to acknowledge the role of Indigenous peoples in stewarding the land on which that event is taking place[6.19]. What rituals or routines could we introduce that signal that we are simply the current group of people stewarding this place, and that there were plenty before us and there will be plenty after? Celebrating the previous and future stewards of a piece of land reinforces our interconnection with past and future generations.

Routines that recognise cycles: We recognise some cycles in our patterns of work. For example, the monthly invoicing cycle and the tax cycle. But there are many living cycles that we tend to ignore in our patterns of professional behaviour. What if we started to work with these living cycles rather than ignoring or flattening them? What if we acknowledged that the daily hours of light and dark change during the day? What if we acknowledged the role that the menstrual cycle plays for many of us? What if we acknowledged the seasons — that there is a time to work and a time to rest and that this pattern will change on an annual cycle? Recognising these cycles enables us to value the abundance in them as well as the need for recovery in living systems.

By shifting routines and rituals we gradually shift those repeating processes that continually reinforce mindsets.

6.3.1.4 Symbols (imagery)

The fourth lens is that of symbols, or as we prefer to think of it, imagery — the visual world of our work. What we see when we walk into our offices. The visual language we use in our branding. The visual language we use in our designs. On our websites. In our submissions when trying to win jobs.

So, what is the visual world of the structural engineer? Walk into many structural engineering practices and we tend to see pictures of completed buildings and in the larger offices, not a lot of greenery!

The images we use are another powerful signal of what we hold important, and also have the power to influence those values. So how might we use imagery to reinforce a collective mindset of interdependence, abundance and emergence?

Show images of the whole supply chain: What if we showed images of the places our waste materials come from rather than the buildings they end up in (Figure 6.16)? And what if we showed images of the places our materials end up?

Figure 6.16: Images of our supply chain might not be pretty but they keep in mind the unseen impacts of our work

Identify patterns from the local system: Returning to the four observational perspectives of ecosystem, materials, community and vernacular (Chapter 4), what are the local patterns that we can show, celebrate and help to integrate into our creative system?

Bring the living world into work: Can we flip our inert offices into thriving, green places? Working among more plants is not just good for us, it is another reminder of our connection to, and dependence on, wider living systems.

The visual cues with which we surround ourselves are like a subtle daily reminder of things to think about. They also signal to others the types of work we want to do. By carefully choosing these visual narratives we can emphasise our interdependence with the places we work, the abundance in the systems we work with, and the opportunity to work with the emerging patterns of place.

6.3.1.5 Organisational structure

Next, we consider organisational structure — the org chart for our work. The organisational structure informs who we talk to and the nature of that relationship, including the inherent power dynamics.

The typical organisational structures and relationships a structural engineer might encounter are:

- A formal hierarchy of managers and subordinates in their organisations
- A network of contributing parties in a design team
- Relationships with a growing number of specialists rather than generalists

Organisational structures and relationships such as these (Figure 6.17) typically emerge to drive efficiency, but can also embed control and division. To create a culture of interdependence, abundance and emergence, changing how we organise ourselves is a powerful tool:

Inclusive work environments: Creating inclusive working environments is fundamental to reinforcing our interdependence; recognising the abundance of diverse knowledge and experiences.

Local expertise: As we become more interested in the specifics of the places we are designing in, the expertise we need may not come from traditional design professionals.

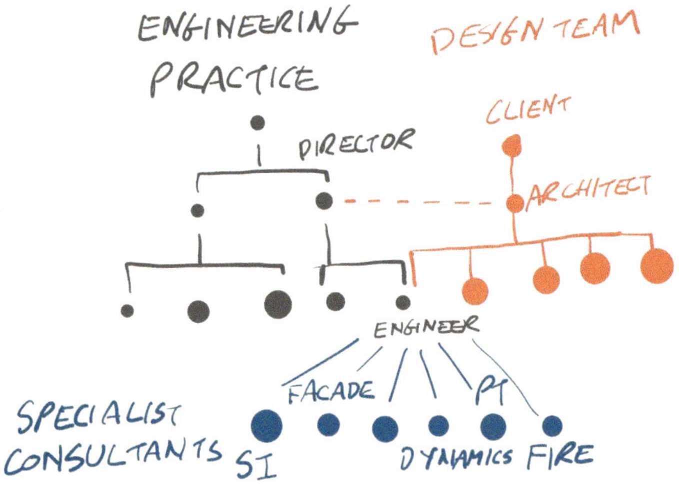

Figure 6.17: A traditional hierarchical project structure

How can we organise ourselves to be able to collaborate with the people that hold local knowledge?

Ecosystem-centric teams: In a fully regenerative approach we would see the ecosystem as part of the design team. So how do we begin to transition to that approach? An organisation that Oliver knows is experimenting with giving the environment a seat on the company's board. To ensure design teams are thinking about their interdependence, how about bringing two empty chairs into design team meetings: one for people and one for the ecosystem in the place we are working?

Physical connection to place: The distance we are from site is a sign of the strength of that relationship. In Chapter 4 we talked about the importance of continued observation in continuous, placed-based design. How we physically connect to the places we are designing for becomes just as important a consideration as how we connect to our colleagues.

These suggestions are intended to highlight how thinking about organisational structure can shape the mindsets that we hold (Figure 6.18).

Figure 6.18: A collaborative project structure built on regenerative practice

6.3.1.6 Power structures (use of power)

The final lens is that of power: How do those who hold power in the organisation use it to reinforce or undermine the culture we are trying to create?

We can think about how power is wielded in structural engineering practices. In the UK, many engineering companies are shareholder owned; at the same time, there is a strong tradition of various forms of employee-ownership or employee-control of engineering practices. But this lens can also relate to the power wielded over the communities and places we are working in. A design project with a good community consultation process might make people in the community feel empowered; but equally, a community that is overruled in planning permission decisions may feel disempowered.

What therefore might be ways that power can be used to create mindsets of interdependence, abundance and emergence?

'Power over' versus 'power to': This concept seeks to differentiate those that use their power to control a situation versus those that use their power to enable someone or a group of people to do something. It is a framing for trying to judge whether someone with power is trying to shut down a situation or open it up.

Give power away: Ultimately, the emergence mindset is about giving power away and working instead with what a system is trying to do. As we have said before, there will always be a role for structural engineers to bring their multitude of skills and knowledge into design, but always with the aim of creating greater empowerment in the communities we are working with — to continue to adapt and co-evolve with environmental circumstances as they change.

When James worked in industry at Integral Engineering Design, there would be an annual 'company away day'. The directors, rather than running the day, employed a facilitator, who was encouraged to give everyone an equal say as they discussed the business.

Power aligned with goals: Many of you reading this book are likely to be employees of companies that have signed a climate and biodiversity emergency declaration. These are strong statements of intent signed by the boards of your organisations. But the true potency of these statements is determined by whether those with power in our organisations back up these statements with actions, or if their actions in fact undermine these declarations. A commitment to regenerative practice backed by the actions of those with power has the potential to really showcase what is possible.

Be aware of your agency: Some of you will believe that you are powerless, that you have no sway on the company that you work for and decide not to try and make a change, rather than try and fail. If you are in this situation don't give up your agency so easily. From your first interactions with a company (probably at the interview stage), to your parting as you move on to other things, you can always politely ask about policies on biodiversity, or inclusivity, or net zero. You never know, people may be more interested in making the changes than you imagine. Often people want to change — they just need a reminder, or a push, or for someone to lead the way.

Equally, some of you will be in positions where "what you say, goes". Wield your power carefully, being mindful of the aforementioned points. But wherever you sit, from graduate to CEO, try and acknowledge the power you have, and use it to create interdependence, abundance and emergence.

6.3.2 Summary

We can use organisational culture to create mindsets of interdependence, abundance and emergence (Table 6.1).

Table 6.1: Questions we could be asking at the mindsets level

Mindsets–level change	Questions to ask
Control systems	• Where do we draw the red line around the site? • How do we move from control to care? • In what ways can we create an abundance of knowledge? • Can we work in a more agile, adaptive and responsive way?
Stories	• How can people tell stories of interconnection, abundance, emergence and transition?
Rituals and routines	• Can we create routines that value biodiversity and that work with the cycles of the living world? • How can land and place be celebrated by rituals that we build into our work?
Symbols (imagery)	• What is the visual world of the structural engineer? • How might we use imagery to reinforce a collective mindset of interdependence, abundance and emergence? • How do the images we use reflect our priorities, and how can they influence our values?
Organisational structure	• How can we shape design teams that are inclusive, that work as part of community and that involve the ecosystem?
Power structures (use of power)	• How is power used within the organisation? • What is the distinction between using power to control a situation versus enabling others? • How can individuals recognise and utilise their power to drive change?

6.4 CONCLUSION

At the start of this chapter we asked: How do we create the mindset for a transition to a more regenerative built environment industry? We proposed three key mindsets: interdependence, abundance and emergence — drawn from the Living Systems Blueprint introduced in Chapter 3. These three mindsets collectively provide a lens through which we can perceive the potential for regenerative design and identify the stepping stones necessary for transitioning to a regenerative built environment sector:

- **Interdependence** emphasises our connection and reliance on the complex human and living systems that we are part of
- **Abundance** helps us to see the potential for the richness that can arise from symbiotic systems and working with what is plentiful
- **Emergence** enables us to see the benefit of working with what a place/an ecosystem/a community is trying to do, rather than what we want it to be

As demonstrated by Case study 25 (the ongoing transition to decarbonise our industry) we witnessed the influential role of groups of engineers in shifting mindsets, thereby unlocking the potential for industry-wide change.

Finally, we explored how culture change serves as a powerful tool for shifting collective mindsets, suggesting various practicable steps at the company level to drive this transformation.

REFERENCES

6.1 Lent J. *The Patterning Instinct: A Cultural History of Humanity's Search for Meaning.* New York: Prometheus Books, 2017.

6.2 Wahl DC. *Designing Regenerative Cultures.* Bridport: Triarchy Press, 2016.

6.3 Wilkinson R and Pickett K. *The Spirit Level: Why Equality is Better for Everyone.* London: Penguin, 2010.

6.4 Johnson R. *Embodied Social Justice* (2nd edition). Abingdon: Routledge, 2023.

6.5 Fleming D. *Lean Logic.* Available at: https://leanlogic.online/ [Accessed: December 2023].

6.6 The Institution of Structural Engineers. *Design for zero.* London: IStructE Ltd, 2021.

6.7 Grist E et al. 'The environmental credentials of hydraulic lime-pozzolan concretes'. *Journal of cleaner production.* 2015:93. Available at: https://doi.org/10.1016/j.jclepro.2015.01.047 [Accessed: December 2023].

6.8 Hammond G and Jones C. *Embodied Carbon: The Inventory of Carbon and Energy (ICE).* 2011. Available at: greenbuildingencyclopaedia/wp-content/uploads/2014/07/Full-BSRIA-ICE-guide.pdf [Accessed: December 2023].

6.9 Roynon J. *Embodied carbon: structural sensitivity study.* Available at: www.istructe.org/resources/case-study/embodied-carbon-structural-sensitivity-study/ [Accessed: December 2023].

6.10 Gibbons O and Orr J. *How to calculate embodied carbon.* London: IStructE Ltd; 2020.

6.11 Smith and Wallwork. *Smith and Wallwork 2023 materials and carbon report.* Available at: www.smithandwallwork.com/whats-new/6234/ [Accessed: December 2023].

6.12 IStructE and Elliott Wood. *The Structural carbon tool – version 2.* Available at: www.istructe.org/resources/guidance/the-structural-carbon-tool/ [Accessed: December 2023].

6.13 Arnold W et al. 'Setting carbon targets: an introduction to the proposed SCORS rating scheme'. *The Structural Engineer.* 2020:(98)10. Available at: https://doi.org/10.56330/SQDI8782 [Accessed: December 2023].

6.14 GIRI. *Get It Right Initiative.* Available at: https://getitright.uk.com/ [Accessed: December 2023].
6.15 Laloux F. *Reinventing Organizations: A Guide to Creating Organizations Inspired by the Next Stage in Human Consciousness.* Brussels: Nelson Parker, 2014.
6.16 MindTools. *The Cultural Web.* Available at: www.mindtools.com/a8im94b/the-cultural-web [Accessed: December 2023].
6.17 The Institution of Structural Engineers. *Circular economy and reuse: guidance for designers.* London: IStructE Ltd, 2023.
6.18 Bristol Institute for Learning and Teaching (BILT). *Publishing a book in a year.* Available at: https://bilt.online/category/publishing-a-book-in-a-year/ [Accessed: December 2023].
6.19 PCMA. *Acknowledging Indigenous Peoples in a Purposeful Way.* Available at: www.pcma.org/indigenous-peoples-land-acknowledgements-purposeful/ [Accessed: December 2023].

PART 2 SUMMARY

This marks the conclusion of Part 2, which has focused on the ways structural engineers can facilitate the transition towards a regenerative built environment sector. In Chapter 4, we discussed the important role of design in this transition. Chapter 5 shifted our focus to making change at the operations level of the construction industry. In this chapter, we have delved into the transformative power of mindsets.

In the final part, we will look at what the regenerative structural engineer needs to sustain themselves in their work, before drawing together all the threads in this book and defining the role of the regenerative structural engineer.

Part 3
Becoming regenerative

In this final part of the book we look at how we can take all we have learnt about regenerative practice — operating at different levels of the system and the different actions we can take — and put them into action using a systematic approach. In short, how can we become regenerative structural engineers?

CHAPTER 7

Sustaining regenerative practice

For regenerative practice to achieve its goal, **for human and living systems to survive, thrive and co-evolve,** we cannot carry out regenerative design on a project-by-project basis. We need to be moving all of our projects towards this regenerative goal. As we described in Chapter 3, we are using Three Horizon thinking to make this transition, and at this stage (where the information and rules at the operations level, and the mindsets of most of those we will collaborate with are still focused on the current paradigm) enabling even small projects is challenging. Which brings us to an important question: **How do we sustain regenerative practice?**

Most books on structural engineering focus on the doing of engineering. How to design a beam, how to detail a column. But this book is different. The focus is on the engineer, not just what they do. In this chapter we will explore three key areas: sustaining motivation, sustaining wellbeing and sustaining imagination. Few books on structural engineering that we are aware of consider personal motivation and wellbeing. But, given that we are talking about a more holistic approach to structural engineering, it would be remiss not to at least touch on these topics. So let's stretch the norms of what goes into a structural engineering textbook and briefly consider how the regenerative structural engineer can thrive.

7.1 SUSTAINING MOTIVATION

We believe structural engineers generally act with good intentions. They are not driven by a desire to make lots of money. They are not driven by ego. They want to design buildings, bridges and other structures that are safe, get built (i.e., are economical, constructable etc.) and function well.

And yet the design we do isn't, overall, doing good — it's doing harm.

To create a transition to a regenerative industry is going to be hard. We know this because of the many times we have tried to make change (and often failed) and we know others who have tried to make change (and often failed). So, what motivates us to keep going in the face of challenges and setbacks?

In this section we will discuss how our values, our sense of personal identity and our perception of agency can motivate the work of the regenerative structural engineer.

7.1.1 Values

There are many reasons why we might want to do the work to help build a more regenerative construction sector. Maybe it is linked to our belief system. Maybe it is linked to our view of what is right. Maybe we believe it makes rational sense. Maybe it is all of them. Ultimately it comes down to our value system.

While we hold divergent religious beliefs (James is a practising Christian, Oliver is a spiritual agnostic) we share the same values. We have the same goal when it comes to the future of humankind and the rest of our ecosystem.

That is why, week in, week out, we have come together to debate regenerative design. Because we value it. In fact, it brings us joy to discuss it with each other. We find it restorative and refreshing. Even when we don't agree. Especially when we don't agree (Figure 7.1).

Figure 7.1: Oliver and James debating the content of the book

Our values ground us and give us direction. Most mornings James has a moment of quiet and stillness. In this moment one of the things he does is to recentre himself and remind himself of his values, ready for the day.

Our values also give us integrity when challenging others in the area of design and construction. Or climate change. Or consumption and loss of biodiversity. Our values can serve as a compass in these challenging dialogues.

7.1.2 Identity

Once we know where we are going it is also useful to know who we are. Really. We are all structural engineers. That is a profound and beautiful part of our identity, but it is not everything. Understanding who we are, how we connect, what our strengths and weaknesses are. Where we need to grow. These all help us to be holistic in our approach to work.

If you are good at listening and observing, but currently spend all day calculating, the world is missing out on the abundance you can offer. If you are good at connecting people, but spend all day drafting, we lose out on the interconnections you could offer. If you are incredible at solving complex statistical problems but are carrying out codified design you could be missing opportunities to use your skills to enable novel material use to emerge. By better understanding yourself you can better support a regenerative system. And to better understand yourself, the 'continuous, placed-based design' model (Chapter 4) is ideal; try new things, gather feedback, reflect, create new opportunities.

Overall, our identity doesn't merely refer to our professional role. It's a more holistic understanding of everything we bring to our work. Working in line with rather than against this identity can enable us to flourish in our work while bring the best of ourselves to the challenge of making change.

7.1.3 Agency

Ultimately, we need to understand that we can make a difference. We need to understand and embrace our agency.

Greta Thunberg is an incredible example of someone who understands their agency. Her book *No one is too small to make a difference*[7.1] demonstrates repeatedly the agency that she carries. And perhaps we too can make a difference. It's about realising that we are not powerless bystanders but key agents capable of creating transformation. By channelling our values, our identity and our myriad skills we can keep finding the agency we need to make change.

Motivation, drawn from values, identity and agency is one way to sustain our work as regenerative structural engineers.

7.2 SUSTAINING WELLBEING

In the next part of our discussion on how the regenerative structural engineer can maintain momentum in their work, we consider wellbeing.

Instead of seeking to summarise the literature on wellbeing into a few paragraphs, we will use the Living Systems Blueprint as a prompt for thinking about wellbeing. In particular, the relevance of interconnection, symbiosis and the capacity to deal with change.

7.2.1 Interconnection

The high levels of interconnection in well-functioning systems creates feedback loops that enhance the system's resilience. The connections between one part of the system and another enable signals to be transmitted that change our courses of action.

When considering wellbeing and interconnection, we are reminded of the importance of paying attention to the signals we receive from our own bodies.

Our body is full of feedback mechanisms, but they only work if we pay attention to them. Gut instinct is trying to tell us something. That yawn is a signal. Just as we need to spend time learning to observe and understand the places, communities and living systems we work in, we also need to spend time understanding and listening to ourselves. Whether through self-reflection, a good conversation with a trusted friend, or through the support of a councillor or coach, it is as important to listen inwardly as it is to observe outwardly.

James has personal experience of what happens when he ignored his body's signals. He ended up in hospital due to a stress-induced anxiety attack. Now, as he describes, he's learning to listen to his body's feedback to hopefully avoid a recurrence. This has included seeing a counsellor, undergoing a six-month coaching course, and building in daily time for self-reflection.

This heightened awareness and commitment to self-care, reinforcing the connections and feedback loops that exist inside us, are essential components of maintaining wellbeing.

7.2.2 Symbiosis

Thinking about how living systems renew their sources of energy can be extremely valuable. A good example is sleep.

Reading Matthew Walker's book *Why We Sleep*[7.2] you quickly realise that there isn't a function in the body that doesn't benefit from sleep. Of the many potential benefits, let's just focus on the ability to have ideas. When we sleep, the N-REM sleep cycle purges our brain of duplicate memories. The REM sleep cycle then reviews all of these memories and reinforces connections between memories that will help us solve a problem we are working on. We find this quite incredible. Effectively, as we are sleeping, the process is not only tidying up the Lego I left lying around the bedroom floor of my brain when I went to sleep, it is sticking together some of the fragments to create something a bit better for me to look at when I wake up.

We can think of the capacity to think, analyse and have ideas as a renewable resource that is recharged by sleep. When we start to think about it, our body is full of renewable resources that are easy to take for granted — our immune system, our ability to heal, our mental health — which we would quickly notice if they broke down. Thinking regeneratively on a personal level is to develop practices that grow these renewable resources.

There are many other examples:

- Spending time in nature observing — which can also be restorative (Figure 7.2)
- Discussing and debating regenerative ideas — which can also be inspiring
- Reading a book to learn more — which can also be relaxing
- Using waste materials to build a den — which can also foster creativity

There are many opportunities for symbiosis at both a personal and community level.

Figure 7.2: To help him slow down and observe the living world around him, James has started to regularly make coffee outside

7.2.3 Responding to change

Here, we consider our ability to respond to change. As we have seen, living systems can self-organise to adapt to changes in their environment.

So how can we increase our capacity to deal with change? Well, one approach is make sure that we are continuously learning. Continuous learning is analogous to the continuous, place-based design process:

- **Observation:** We begin by observing the world around us and how we are responding to it
- **Brief:** We can then reflect on what new skills, practices or knowledge we need in order to respond to it
- **Ideas:** We gather new ideas, ways of thinking and doing things from books, training, courses, meeting new people etc. (Figure 7.3)
- **Testing:** We put these new ideas into practice and see what is working

And the process repeats with observation once more.

Figure 7.3: To enable us to write this book we have been on a continuous learning cycle, reading a number of books (Appendix D)

We see from the Three Horizons model that the norms we take for granted will give way to new ways of doing things. We can sense from the constant supply of news we receive that the world is changing around us. However, if we feel equipped to change along with it, this process can foster growth and enable us to thrive.

Bringing all these threads on wellbeing together:

- **Interconnection** encourages us to acknowledge and respond to our body's feedback mechanisms, enhancing our personal resilience
- **Symbiosis** encourages us to think about cultivating practices that enhance our renewable internal resources e.g., our sleep, immune system and mental health
- **Capacity to deal with change** encourages the adoption of a mindset of continuous learning in response to the changing world around us

Together, these approaches to personal wellbeing can help sustain us in our work as regenerative structural engineers, as well as providing a direct analogy for the change we are trying to make through our work.

7.3 SUSTAINING IMAGINATION

Finally, let's turn to the role of imagination in sustaining the regenerative structural engineer. As Rob Hopkins describes in his excellent book *From What Is to What If*[7.3], the climate crisis is a crisis of the imagination. So many of our ways of being and working are locked in by the paradigm that surrounds us, that it becomes hard to imagine an alternative reality. When there doesn't feel like there can be any alternative, we can become despondent and disengaged.

So how can we fight off this crisis of imagination? Again, one final time, we turn to our Living Systems Blueprint for inspiration.

7.3.1 Inspiration from interconnection

There has never been a time at which we have been more globally connected to people and knowledge. While our supply chains are likely to become more localised in a regenerative industry, there is no reason why we shouldn't seek global wisdom to our problems.

There are two directions we can take with this. We can explore practices from all over the world. Many of our case studies are international. They show a different way of doing things that we could use in our context. We can use our professional networks to access information from almost anywhere. Showing what is possible in one place may be just the trigger to challenge what you are doing where you live.

But we don't just need to seek global wisdom from afar; there is a wealth of global knowledge in our communities. You can talk to different people on your street and you may learn about how climate is affecting different countries. These conversations make that global-local connection.

Local conversations with different parts of our communities can also help everyone tune into the social injustice that is built-in and that we need to be addressing if we are to create greater socio-ecological wholeness.

As engineers we can speak to community groups to find out more about both the challenges that people in our local system are facing and the active and latent capabilities that exist in the people that surround us. And this could be the trigger for imagining: What if the construction industry could be a force for a good in the places that we live?

7.3.2 Inspiration from abundance

The most popular creative thinking technique in Oliver's conceptual design training for structural engineers — measured by volume and enthusiasm — is called 'Ask what if?'

It works by tricking our brain to approach a problem from a different perspective or frame. When we change our thinking frame, we let in new possibilities that our brain might otherwise have filtered out as being impossible. From this new perspective we may discover things we previously dismissed as impossible, but which may turn out to be possible.

To facilitate this process, we ask engineers thinking about a design brief to shift an aspect of the brief to make it sound ridiculous. The bigger the change the better. Then, for a few minutes, they brainstorm responses to this ridiculous brief. Finally, they return to the original situation and see if any of their unlocked thinking has provided a new way of seeing how to proceed.

The connection we make to regenerative design is to see what happens when we ask: 'What if?' from the perspective of an abundance mindset. Through this frame:

- We see the world with generosity
- We focus on people's willingness to give rather than fearing what they may take
- We assume that we will have enough, rather than fear that we will have to do without
- We see the potential of what people could do rather than the risk of what they can't
- We trust to emergent solutions rather than seek to control things out of fear

By adopting an abundance mindset and asking: 'What if?', we can keep rediscovering the potential in the places we live in and work, and this curiosity can keep challenging the status quo.

7.3.3 Inspiration from emergence

Sometimes, to have ideas, we need to let a situation play out and see what happens. The operative word is 'play'. Play is one of our best tools for learning and problem solving. But play is undervalued. As John Wright says in *Why is that so funny?*:

> "We live in a rational and literal culture that doesn't value the frivolous, the pointless, the meaningless. These go against everything we've ever been taught... but if the work is playful it becomes pleasurable, and when you're enjoying yourself you take more risks. Choices proliferate. Problems become more manageable because our perspectives change once the work becomes a pleasure."[7.4]

Play has lots to offer in sustaining the imagination of the regenerative structural engineer.

Play serves as a way to help us imagine what is possible. Play creates a vehicle for co-creation of ideas. Play helps to build trust between people. Play challenges conventional wisdom and makes space for new ideas to emerge.

One technique we have found to be helpful in the Regenerative Design Lab[7.5] is creating an 'artefact from the future'. It could be a magazine article from the future, or design for a building in 20 years' time, or the acting out of a scenario.

Chapter 9 of this book is an epilogue; an 'artefact from the future', created by James. It is his imagining of a day in the life of a structural engineer in 2050. When he first wrote this piece, he read it out to a group of 18 built environment professionals as part of a wandering tour of a woodland. James read a section and then we wandered and then he read the next part. This meandering narrative journey led us out of our usual ways of thinking and prompted us to imagine how different the world could look and what our role in it might be (Figure 7.4).

Figure 7.4: Playfully exploring the future of engineering

Clearly, an artefact from the future is just one person's view, but it is the act of creating and sharing this thinking with others (and this itself giving rise to further creative dialogue) that is interesting. When we connect with others, when we stop and ask: 'What if?', when we open ourselves to seeing what emerges from playing with ideas, we unlock the power of imagination to drive us forward in our work.

However you decide to do it, the important thing is to keep imagining. To keep on asking questions:

- What if the act of building brought people and the living world closer together?
- What if we generated construction materials rather than extracted them?
- What if our built environment could constantly adapt and respond to our changing environment so that we and the rest of the living world could survive and thrive within our ecosystem limits?

We are confident that continuing to ask questions like these, and playing with the ideas they inspire, can help sustain you in your work as a regenerative structural engineer.

CONCLUSION

This chapter explores the idea that structural engineers will need strategies to sustain themselves if they are to play their part in the challenging work of transitioning the construction industry to regenerative practice. This focus on maintaining oneself is consistent with the holistic philosophy that regenerative design supports. We considered three factors to sustain us in our work.

At the heart of our role as regenerative structural engineers is our intrinsic motivation — drawn from our values, sense of identity and agency. It's this personal inner drive that has an important role to play in moving us forward.

We drew inspiration from the Living Systems Blueprint to see how interconnection can inform our wellbeing. This includes listening to feedback from our own bodies as well as from the living systems around us.

Finally, we drew on Rob Hopkins' idea that the climate crisis is really a crisis of the imagination. But with the vast wealth of information at our disposal — both local and global — we have great wisdom to draw on that can seed our ideas. The idea of an abundance mindset can help us see new opportunities, and open us up to new possibilities. And while our purpose is serious, we need to maintain a playful and curious approach to be able to see what we could do together.

The journey of the regenerative structural engineer is not just about traditional 'professional development' — it weaves in values, wellbeing and imagination to make sure we are bringing our whole selves to the journey.

In the next chapter, we draw together everything we have explored in this book to ask: **What is the role of the regenerative structural engineer?**

REFERENCES

7.1 Thunberg G. *No one is too small to make a difference.* London: Penguin, 2021.

7.2 Walker M. *Why we sleep: The New Science of Sleep and Dreams.* London: Penguin, 2018.

7.3 Hopkins R. *From What Is to What If: Unleashing the Power of Imagination to Create the Future We Want.* White River Junction (Vermont): Chelsea Green Publishing, 2019.

7.4 Wright J. *Why Is That So Funny?: A Practical Exploration of Physical Comedy.* London: Nick Hearn Books, 2006.

7.5 Constructivist. *Regenerative Design Lab.* Available at: https://constructivist.co.uk/regenerative-design-lab/ [Accessed: December 2023].

CHAPTER 8

The role of the regenerative structural engineer

In this penultimate chapter we ask: **What is the role of the regenerative structural engineer?**

To answer this question, we take a final journey through the Systems Bookcase (Figure 8.1).

We begin our journey by returning right to the top of the Systems Bookcase. In Chapter 2 we saw the power of paradigms and goals to shape the outcomes of a system.

Figure 8.1: Combining the Three Horizons and the Systems Bookcase models

The climate and biodiversity emergency are showing us that we need a new paradigm for our industry. One that shifts us away from continuous growth to one that enables the thriving of all life, not just human, in the places touched by our work.

This paradigm is called 'holism'. Within this holistic paradigm regenerative design is a systems goal. The goal of regenerative design is **for human and living systems to survive, thrive and co-evolve.**

The central idea in Chapter 3 is that we can't simply step straight from our current practices to fully regenerative practices across our sector. We need to create a transition towards them. In this light, we describe the role of the regenerative structural engineer as collaborating with others to create the transition towards a regenerative built environment industry. Our main challenge is to ensure that our efforts don't just reinforce the existing paradigm (H2- interventions), but start to stretch it towards a more regenerative approach (H2+ interventions).

To guide us, we introduced the Living Systems Blueprint for Regenerative Design, which highlights three characteristics of thriving living systems:

- Interconnection
- Symbiosis
- The capacity to self-organise and adapt

Weaving these characteristics together, help us understand how a regenerative built environment system might be organised. This blueprint serves as our guide, helping us assess whether our actions are truly paving the way towards a regenerative future or merely extending the existing paradigm.

A theme running right through this book is viewing the construction industry as a system, and one in which we can intervene at different levels. So, having recognised that we need to help the construction industry transition towards more regenerative practice, we have looked at three levels at which a regenerative structural engineer can intervene: design, operations and mindsets. Let's recap each of these in turn.

8.1 INTERVENTION THROUGH DESIGN

The decisions we make through design impact significantly beyond the site on which we are designing, stretching far back to the source of our materials and into the future legacy of our construction. The decisions we make in design also have the opportunity to influence the way the construction industry operates, more widely. To leverage the power of design we introduced the concept of continuous, place-based design, with four cyclical stages:

1. **Observation**

   Design work should begin with a period of observation and engagement — and we are likely to spend longer in this stage, than we have previously. Through observation we can increase our interconnection with site, stakeholders and the broader living world. We can identify potential for symbiosis and abundant sources of materials. And we can start to notice ways to intervene that relate to what the local system is trying to do.

   We suggested four areas to focus on in observation: communities, ecosystem, materials and local construction processes.

   Through deep observation we extend our understanding of the impacts of our work far beyond the traditional boundaries of a construction site. The richer our understanding, the greater our capacity to create designs that are embedded in (and are an extension of) the existing social and living systems.

Through observation we also recognise the complexity that comes with our extended supply chains. Thus, transitioning to a more regenerative built environment industry involves bringing more of that supply chain into view so that we can more fully assess the impact of our decisions.

2. **Brief development and interpretation**
   We see this phase as the most powerful opportunity we have, to influence the impacts of a project. We introduced four principles for leveraging the power of the design brief:

   The first is to aim higher in the system. This step involves shifting from thinking about designing standalone buildings towards seeing design of buildings and infrastructure as a catalyst for wider systemic change.

   The second principle is to aim to re-establish interconnection. This challenges us to redefine the traditional parameters of design and create a more interconnected approach.

   The third principle at the brief stage is to aim to build symbiosis; identifying opportunities for creating, enhancing or unlocking symbiotic and abundant systems of material supply.

   The fourth principle is to enhance the capacity to co-evolve. This shifts our thinking away from doing things to the world, towards working with the potential of a specific place to create the best-fit design for the local human and living systems.

3. **Idea generation**
   The third stage in continuous, place-based design is idea generation. Here we use the Creative Systems Model to think about what it would mean for the communities of people and the living systems we are working with, to have ideas together.

Through taking on roles as creative facilitator, connector, evaluator of ideas and educator, the regenerative structural engineer can support idea generation that is more strongly based on emergence.

4. **Making and testing**
   The fourth stage in the continuous, place-based design process is making and testing. The first point here is to define what we are testing for, and in this we can use the Living Systems Blueprint as our guide: looking to build interconnection, symbiosis and the capacity for self-organisation. The second point is how we test. In this, we recognise that to respond to the complexity of human and living systems, design needs to become a continuous process of observation, adjusting the brief, developing ideas and re-testing.

   Each stage in the continuous, place-based design process provides an opportunity for transitioning towards a more regenerative construction industry.

8.2 INTERVENTION THROUGH OPERATIONS

While design presents us with many opportunities to transition towards regenerative practice, what we can achieve is ultimately limited by the rules, regulations and procedures that sit on the operations shelf of the Systems Bookcase. But here too, we can find opportunities for the regenerative structural engineer to help create a transition towards more regenerative practice. We defined five operational contexts in the which we can find opportunities to intervene:

1. **Place of work:** This context is about looking at the systems we use to manage our projects, the relationships we form and the environment we create to work in.

2. **Collaborators in industry:** This context is about building the wider group of people with whom we can create change in our industry. What are the barriers to wider collaboration and what is the potential that could be unlocked through these relationships?
3. **Supply chain dynamics:** This context is about removing the industry-level blockers in the supply chain, to transition to more regenerative practice, which accounts for the full impact of supplying materials, not just the impact at point of use.
4. **Civic sphere:** In this sphere we have the opportunity to influence planning decisions around shared resources, and can find opportunities for the design and construction process to be more empowering for the people that live in the places we work with.
5. **Wider industry:** This sphere includes all the organisations and bodies that drive innovation and regulate practice. These organisations have significant influence, but they are all made up of people, and we can help to influence their decisions in a variety of ways; from supplying case studies and information, to writing letters and joining committees. There are a number of ways we can impact wider industry.

To intervene effectively at systems level, we need to identify and understand the underlying principles (known as 'causal loops' in systems thinking) that drive the behaviour we see, and the points at which we may be able to initiate change.

8.3 INTERVENTION THROUGH MINDSETS

The third level of intervention for the regenerative structural engineer is that of mindsets. Our mindsets influence the way we see the world, what we may fear and what we may hope for. Mindsets have the power to limit our thinking or open up new possibilities.

We propose three shifts that could help catalyse a transition towards a regenerative built-environment sector.

1. **From separation to interdependence:** In this mindset shift we aim to see ourselves as not just connected to but dependent on the health and wellbeing of the human and living systems that surround us. This in turn has the potential to drive interest in the 'second site', to cause us to account for embodied biodiversity and embodied social justice as well as embodied carbon and establish a stronger imperative to design for human and living system wellbeing.
2. **From scarcity to abundance:** This second mindset shift builds on the first to enable us to see the potential in the living world to provide us with the resources that we need in a way that encourages all parts of the system to thrive. It also shifts our thinking to seeing how living abundance comes in cycles and that some potential abundance may currently exist in a depleted, desertified state that needs help to recover.
3. **From control to emergence:** This third mindset shift aims to unlock the capacity for living and human systems to co-evolve. From an emergence mindset we might see the opportunities for greater collaboration in our work. We might see the benefit of greater sharing of information. And ultimately, we can see the potential for the best designs to come not by driving forwards an idea, but by working with what is trying to emerge as the best response to the complex needs of a place.

Finally, we went onto explore how organisational culture can be used to reinforce mindsets, and how the regenerative structural engineer can use the levers of cultural change to shift mindsets towards more regenerative thinking. To identify possible courses of action we looked through the six lenses of the Cultural Web[8.1].

1. **Control systems:** This lens considers what we measure and how this signals what we think is important. We consider: the impact of where we draw the red line around a site; the notion of control versus care; intellectual property; agility in our approach. These are just four examples of how our control systems could create mindsets of interdependence, abundance and emergence.
2. **Stories:** This lens acknowledges the powerful role that stories can play in shaping culture. How could we tell stories of interdependence, abundance and emergence? How can we create opportunities for storytelling? By being conscious about the stories we tell, we start to challenge the collective narrative about what we consider to be important.
3. **Rituals and routines:** These are the patterns of behaviour that signal what is acceptable within the collective mindset of the organisation. Examples are routines that value biodiversity and recognise cycles in living systems, and rituals that acknowledge land and place. By shifting routines and rituals we gradually shift those repeating processes that continually reinforce mindsets.
4. **Symbols (imagery):** The images we use and surround ourselves with are a powerful signal of what we consider to be important. Integrating into our work and workplaces images of the supply chain, patterns from the local system and even living things themselves can serve as daily visual cues, reminding us of what we value most.
5. **Organisational structure**: The organisational structure shapes who we interact with and the nature of that relationship. Creating inclusive work environments, working with local experts, creating eco-centric teams and creating a physical connection to the places we are working with, are all ways that organisational structure can help reinforce more regenerative mindsets.

6. **Power structures (use of power):** Through this final lens we look at how power is used to reinforce culture and therefore our mindsets. Do those with power in an organisation use that power to control or enable? Is power given away and distributed or concentrated? Does power reinforce or undermine goals? And finally, do we fully realise our own agency - our own power - to make change?

Overall, the role of the regenerative structural engineer is to:

- Hold and sustain an image of a viable future in which the built environment industry contributes to the flourishing of human and the living world
- To use all their many different skills to intervene at the levels of design, operations and mindsets to help build the transition to that viable world (Figure 8.2)

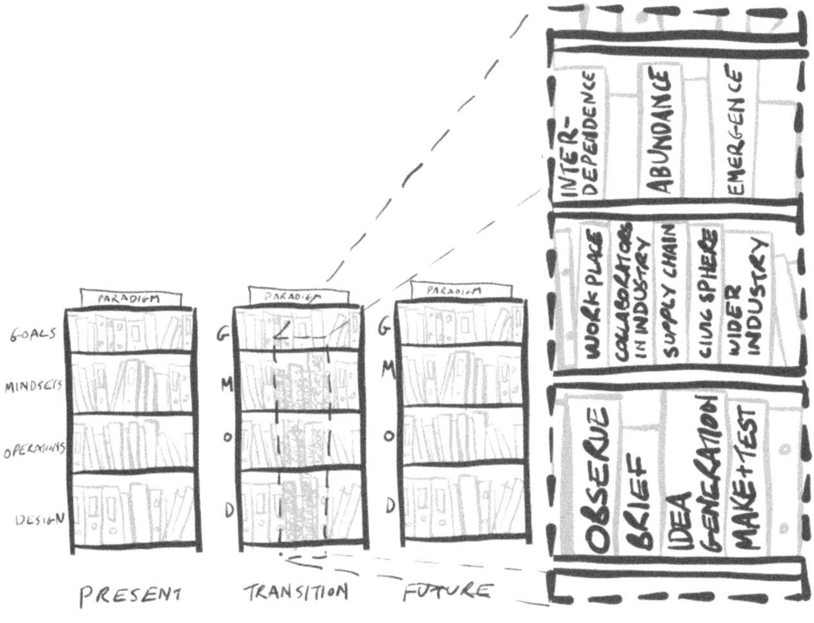

Figure 8.2: Populating the transition bookcase

8.4 THE POWER WE SHARE

Through the course of this book we have come to see the profound influence of paradigms over systems. But while paradigms exert great control over systems, they can be challenged, shifted and eventually transformed.

The questions in Table 6.1 on power remind us of the power that each of us hold. When we see our ways of working as fixed, we may feel powerless. But when we realise the myriad ways we can intervene, we can begin to see our influence. This isn't 'power over' but 'power to' — the power to create a transition to a regenerative built environment sector.

This isn't power that we wield in isolation, it is something that we ignite in others. We won't reach our destination in one giant leap, but through a series of small collaborative steps. The mindsets that a few of us adopt, help us identify the operating rules we want to challenge. The rules we challenge at operations level enable innovation at design level. Demonstrating the difference we can make at the design stage gives those at the operations level more confidence to change the rules of the system. These new rules, in turn, shape the culture and mindsets we adopt.

In this way, we journey up and down the Systems Bookcase; with each step, nudging the construction industry further away from harmful practices and closer to a system that drives positive outcomes in every facet of its operations. Holding this future in mind, the role of the regenerative structural engineer is to actively create that future, one step at a time. So that one day, in a future that is still far over the horizon, we can be confident that when we are designing a building, we are not merely minimising harm to the environment but actively contributing to its healing and regeneration.

REFERENCE

8.1 MindTools. *The Cultural Web*. Available at: www.mindtools.com/a8im94b/the-cultural-web [Accessed: December 2023].

CHAPTER 9

Epilogue: 2050

In the writing of this book, we spent a lot of time thinking about what Horizon Three might look like if we were to practice placed-based continuous design. The thinking has informed much of the content and, as an epilogue, we wanted to present to you this 'artefact from the future'.

The brief for this artefact from the future is to imagine what our role, as regenerative engineers, might look like in the year 2050. To enable us to provide a solution to this question we have made some assumptions. Firstly, we assume the engineer in question is James. We assume that he is based in Bath, UK, in the same house he lives in now, which he has already lived in for the last 20 years. We also assume he is still alive and able to practice as an engineer. These assumptions have enabled us to observe human and living systems, try out ideas, and collect feedback, based both on our own observations and on the feedback of others. While this artefact from the future is limited (it does not give answers to what the role of a regenerative engineer might look like in another place, country or organisation type) we hope it is helpful.

The scenario is intentionally set in an optimistic future, it is the possible future that we are hoping to create[9.1]. We have tried to use the opening chapter of *From What Is to What If*[9.2] as our starting point. As we go through the artefact we reference a number of future-facing projects, but we have also added our own thoughts. We have tried to avoid having too strong a sense of the political and economic outlook of 2050 as we trust that while we may have different views on this, we may share a vision of a regenerative engineering future.

This is not a one-size-fits-all exercise. It is unique to James, his location, his observation, and his experience. We would encourage you, once you have read this epilogue, to create your own vision of 2050. To assist in doing that we have provided some suggestions in how you might approach such a task in Appendix C.

So, let's jump to 2050!

9.1 AT HOME

Welcome to my street. A lot has changed in the 50 years I've lived here. There are now very few cars on the roads[9.3]. Here there are none. Every roof is covered in solar panels, solar tubes, solar tiles, solar paint. Outside of the 'heritage zone' (which is shrinking annually) all houses are now overclad with insulation. Windows at ground floor level have insulated shutters to minimise heat loss in winter and heat gain in summer. Many windows at first floor have either been covered over partially or completely for the same reason (Figure 9.1).

As technology progresses, it stands to reason that solar power becomes more widely available in various formats.

Bath is a World Heritage Site and preserving its historical culture is important. However, the area over which this status is placed may reduce over time as a result of other pressures.

Figure 9.1a: James' street in 2023

Figure 9.1b: How James' street might look in 2050

One of the biggest changes is in the purpose of the loft, which now has considerable water storage. Air and ground source heat pumps require tanks. In addition, we collect rainwater from the roof for a variety of purposes. Some is treated, but most is used for irrigation and flushing. We still have mains water, but it is metered and expensive so mostly we try and utilise what we can gather. To maximise the storage, we have switched many of the floor and roof joists around. It turns out that while most floors in buildings are overdesigned, roofs are typically under designed. We used whatever was available to create the water storage — it's not uncommon to find baths and even pond liners up in the roof. We can no longer afford to fill and heat a bath and it is more efficient to reuse it than pay the recycling charge.

Of course, for a short while people assumed they could increase the amount of storage in a roof without the expertise of an engineer — but when the ceiling joists failed after a heavy storm round the corner, which led to the failure of the first floor and the progressive failure of the house, people changed their minds. I was quite busy for a while just assessing all my neighbours' properties, mostly on weekends and evenings.

We didn't waste any of the collapsed building. As it was a 1950's concrete post and block prefab construction which was designed to have a working life of just 20 years

The history of the buildings in my area has been passed on verbally from residents who moved in to the very first dwellings.

(extended to 80+ years), we weren't able to reuse the materials for any primary structure, but we did use them to create a large communal composting pen[9.1] (the post and block was perfect for this). We also repurposed the timber floor and roof joists and over-cladding that was added later, to create a community composting toilet. While we all have our own toilets, solid waste is actually really valuable if well treated, but only when produced in sufficient quantity. As a result, individual households mostly failed when they tried to create composting toilets, but a community facility has flourished (or more accurately our crops have flourished as a result). Of course not everyone chooses to use it, but enough to make it viable. The addition of good insulation and trace heating from solar panels makes it much more inviting.

Upcycling is not a new idea, but the idea that we should **continually** upcycle is taken from Daniel Wahl's *Designing Regenerative Cultures*[9.4].

As you can see, engineers are busy, whether checking existing joists under increased loading, foraging for parts to make community facilities, or using their broader engineering skills to think of creative ways to optimise what is available for the local community. The aim is to continually upcycle what we have for better outcomes (Figure 9.2).

Figure 9.2: How the city of Bath might look in 2050

9.2 DECONSTRUCTION PROJECT

We are waiting to meet with the site engineer. Before she arrives let me explain. Most projects these days reuse the existing building[9.5]. However, to ensure material is used wisely, if a building has been converted twice in any ten-year period, it can't be converted a third time — it is deemed the building can no longer function as a whole and it needs to be deconstructed[9.6]. The same is true if it has stood empty and unutilised for more than 12 months. This is to ensure all resource is well utilised. This building has just defaulted under the Utilisation Bill and now we need to take it apart in such a way as to maximise the future opportunities. Foundations must be reused. All other materials must be converted into usable parts and shared with the city or region. While some may be reused on the site, we have to demonstrate that the reconstructed building is sufficiently different in configuration to be of use. While this sounds simple there is enough bureaucracy and challenge to make this approach much less attractive than adding the materials to the local inventory of reused components. Here is the lead contractor who will be overseeing the deconstruction work:

"Holly, so good to see you again, how are you?"

"Good. So, what have we got here?"

"Holly, it's not good news I'm afraid. Concrete frame, designed in 2018, constructed in 2022."

"You sure? You know those are my least favourite right?"

"I know. At least it isn't the AI designed and quality-assured buildings of the late 2020s and early 2030s, they are the worst. The AI saw the gradual increase in load, stiffness and interconnection (in the name of robustness) and assumed that buildings should be optimised in this direction. As the designs were black-box they were hard to check, and the initial reduction in material suggested they were improving the design. It took a while for us to realise what was going on. Coupled with this was the automated QA. While QA is a good thing, no account was made in the design for the reduced uncertainty as everything was built to spec. Concrete grades are typically up by $20N/mm^2$ in this era. And the detailing is a nightmare. These buildings are designed to stand for ever! Anyway, sorry for the rant – I'm so glad I wasn't in industry at that time! Back to the issue in hand. We need to think not just about extracting the parts, but upcycling where possible. It's a concrete flat slab, monolithic construction, designed for $5kN/m^2$ everywhere even though it was intended as an office."

The MEICON survey[9.7] shows that offices are regularly designed for $5kN/m^2$ even though only $2.5kN/m^2$ is required by the code.

"What about the stability."

"The stability system is generally logical, so at least the deconstruction won't lead to instability and possible failure — it would be good to avoid a collapse and the associated financial penalties."

Designing for deconstruction doesn't just mean being able to take the structure apart, but being able to do it safely. Considering reversing construction processes so the structure remains stable during deconstruction is also an important consideration, especially if the design information and build sequence is not available.

"Any chance we can take it apart?"

"Sadly not. Although deconstruction was understood and engineers had to provide a deconstruction plan for BREEAM, generally it was ignored. The only way we are taking this one apart is by cutting it into bits."

"Seriously. Have you got all the necessary permits? Despite improvements, cutting still emits far more particulates and unusable waste than allowed without a permit, and the waste is just that, waste — we won't be able to recapture or reuse any of the materials lost in the cuts."

"I know. But until we have found another way to do this through localised dissolving of the binder, we have no other choice. Add it to the citizen research list. I'm sure there must be a solution for deconstructing using non-harmful chemicals with zero waste, we just haven't found it yet."

Part of co-creation is that research questions should come from our community

Use of non-harmful chemicals is a repeated theme in *Designing Regenerative Cultures*[9.4].

"Foundations?"

"Piles and pile caps. We need to reuse them in the next site use."

"Please tell me we are looking at regular grids, fixed using the international standard?. Actually, don't answer that, I already know."

There is currently no international standard for building grids, but if grids were regularised, reusing parts of buildings would become substantially easier/ more material efficient and would enable parts from multiple sites to be combined more easily.

"Sorry, grids are all over the place. This one is going to be a nightmare to reuse. Column positions jump all over the place. We may need to consider converting the pile caps into a monolithic raft. Maybe we can argue for a building of state, and rigid construction, you never know. But for now we just need to map the locations and explore solutions. Sorry it's not better news. Next time it will be an old building, designed with sensible loads and constructed in a way that can be reversed, I promise."

"That's what you always say, but those buildings never seem to come up for deconstruction, it's always these overdesigned monoliths. Like the last one. You would have thought it would be easy to design a steel frame as demountable, but then they bonded everything together with concrete. Disconnecting the slabs from the beams was a total nightmare. How hard would it be to make the stud connection reversible? I wish everyone who designed a new building had to jump in a time machine and deconstruct it. Then they would think differently about the whole thing. Seriously, how hard is it to design for deconstruction? It's not like they were designing the pantheon, it's a department store. I mean how could they not see that its useful life would be less than 30 years with no meaningful opportunity for reuse."

While buildings have a design life of 50 years the useful life can be much longer or, as is the case with some building types, shorter. Specialist buildings with short useful lives are difficult to repurpose.

"Holly, I totally hear you. I wish I had taught design for deconstruction much earlier, but I didn't. We are all learning.

But you're right — if I had that time machine I would have done so many things differently."

9.3 IN THE OFFICE

Welcome to the office. Let me show you round. Firstly, this is a cooperative space. We are not an engineering practice so much as a community body who work collaboratively to try and enhance our local area. We have artists, accountants, architects, ecologists, civil engineers, geologists, designers, systems designers, photographers, journalists and anyone else who is interested in working together. Many processes don't need the input of a structural engineer, but when they do we become a leading member of the team, as we enable the idea to happen. While experimentation is key, and we are no longer worried about serviceability limit states in the same way, structural and ecological safety remain paramount and we work, together with the ecologist, to ensure the project will not only lead to human and living systems flourishing, but also to an acceptable probability of no harm to human life and ecology.

We have learnt to trust each other. To seek opportunity. To celebrate diversity. And to expect ideas from any member of the team as we search for the best way forward.

The space is divided into four. At the front we have our collaboration space. Most buildings avoid the overuse of glass but here the glass is essential to create a connection with our community. The space enables us not just to collaborate inhouse, but also with key stakeholders across the city. People are encouraged to share ideas with us. However, the glass also reminds us that the city is our office. We search for materials. Spend time listening to the ecology and seeking opportunity. No moment is wasted. Every slow walk. Each pause on a bench. Every morning coffee. Each provides feedback, data and understanding of the ecosystem. Design now isn't just collaborative, it is inclusive. Projects need to reflect the needs of the community. They need to be of benefit to as many people as possible, not just a few[9.8]. We have worked hard to build community trust to pursue projects that support the whole community. That trust means that people now share a wealth of information. As a result, we try, as much as possible, to have glass doors open, invite the city to come in as we observe the city as we work.

The second space, which is connected to the first, is our testing lab (Figure 9.3). Here we build prototypes, test material samples from both existing buildings and possible new materials. We are only a small setup and the big

Figure 9.3: Office testing facility

testing programmes happen up at the university, which I continue to visit frequently. This space also talks to the community. It shows that we are constantly innovating. This isn't marketing, its relationship building. I like to think that the community is proud of the work we do here. While non-engineers may struggle to understand our probabilistic analysis and use of advanced computing, the physical act of making and testing is accessible to all. And while we have a strong health and safety culture, as we value everyone who enters the lab, we also look to use this space in a collaborative way. We have a good selection of *in situ* testing equipment, which we regularly lend out to the community. While we could do the work ourselves, time is limited and so we encourage, at least in the first instance, self-diagnosis. Insurance no longer covers structural integrity (and no longer exists in the format it did in the 2020s) and people are left to repair their own buildings. Most issues are due to buildings being overly rigid, which leads to cracking, but sometimes they are more serious. By aiding self-diagnosis we can focus on the safety-critical issues.

Behind the testing space is our archive. This is maintained in both digital and printed formats. While, as engineers and community enhancers, we are a protected organisation and have a higher guarantee of energy than most, power outages do still happen, and can last for a while. As a result, we try and keep the most significant documents in printed form. We have a section on historical approaches to design, which includes a large number of design guides and codes of practice. We also have the most up to date process documents. We keep our own local material inventory. This covers the properties of local grown materials like timber, bamboo, hemp and straw. We have information on local abundant materials such as earth and Bath's famous limestone. We have information on waste stream materials, both local to us and nationally. And we have information on key existing infrastructure including as-built drawings, surveys and all material testing results. Our library of materials information enables us to design much more efficiently than national practices that don't have access to such a wealth of information (although we always share information when asked). We also value analogue information for other reasons; the beneficial way reading a printed book helps restore our own wellbeing, and the value we place on the information due to the precious nature of the artefact are examples.

The final space is the focused analysis lab. This is a space for deep thinking. We have high-end computers for advanced analysis. We no longer run simple linear analysis. Instead we use enhanced processing power to run non-linear analysis to optimise design and if possible we run a full probabilistic analysis. That way we can better understand how likely something is to fail. Where projects have a negative impact (in the form of loss of biodiversity, pollution, emissions etc.) we run a detailed risk analysis to ensure that the risk of failure is optimised against the environmental risks. All projects of this nature must be agreed and signed off by building control and must be in alignment with both Approved Document A (Structure) and X (Ecosystem Impact).

However, the work of a designer is not just the analysis. It is all the thought and process that occurs before the analysis, in fact the analysis really is just the icing on the cake. So, in the space we have both digital and analogue drawing boards (and hybrid, which I am still struggling to use). And each process has a series of drawings which capture the full thought process behind the design, linked to both the digital archive and the stakeholder conversations and analysis. This space also contains the space to dwell. To think deeply. To make systems connections. And to record them in whatever manner is most comfortable, from video to scribbled notes (which I prefer). This dwelling space includes immersive technology. While sitting in an ecosystem is highly beneficial it is also limited by human time. This immersive space enables us to speed up and slow down the cycles of the space and to add our interventions and see their impact. Humans interact with space in our linear time domain, driven by the rhythm of our heart and breathing, the sun and the moon. But as we immerse ourselves in the ecosystem it is important to recognise the different cycles and rhythms. The immersive technology enables us to do this.

As part of my research for the book I visited Format Engineers who discussed how they are using AR to show different stakeholders different design solutions in their actual context. This type of immersive technology could be very powerful.

The BBC documentary series The Green Planet[9.9] used different camera rates to capture different patterns in our ecosystem.

9.3.1 Reading *The Structural Engineer*

Every day, whether I am in the office, working from home, or out on the road (which means travelling by train and then finding local urban transport to get me to my destination — not using an actual road) I always make time to think, dwell, and read. So, let's stop for a coffee. 30 years on and analogue devices still delight me. A hand grinder and hand press for the coffee. A record on a record player. A comfy chair. A good set of wired headphones if I'm on my own. And a physical book or magazine to read.

It's a Tuesday, which is *The Structural Engineer*'s weekly publication day. Demand for innovation and new ideas (coupled with global perspectives on design) increased, and in response, *TSE* increased its frequency to weekly in 2042. Although it's available digitally of course, it's print format is still highly desirable. Let's have a look at the content of this week's issue (Figures 9.4 and 9.5).

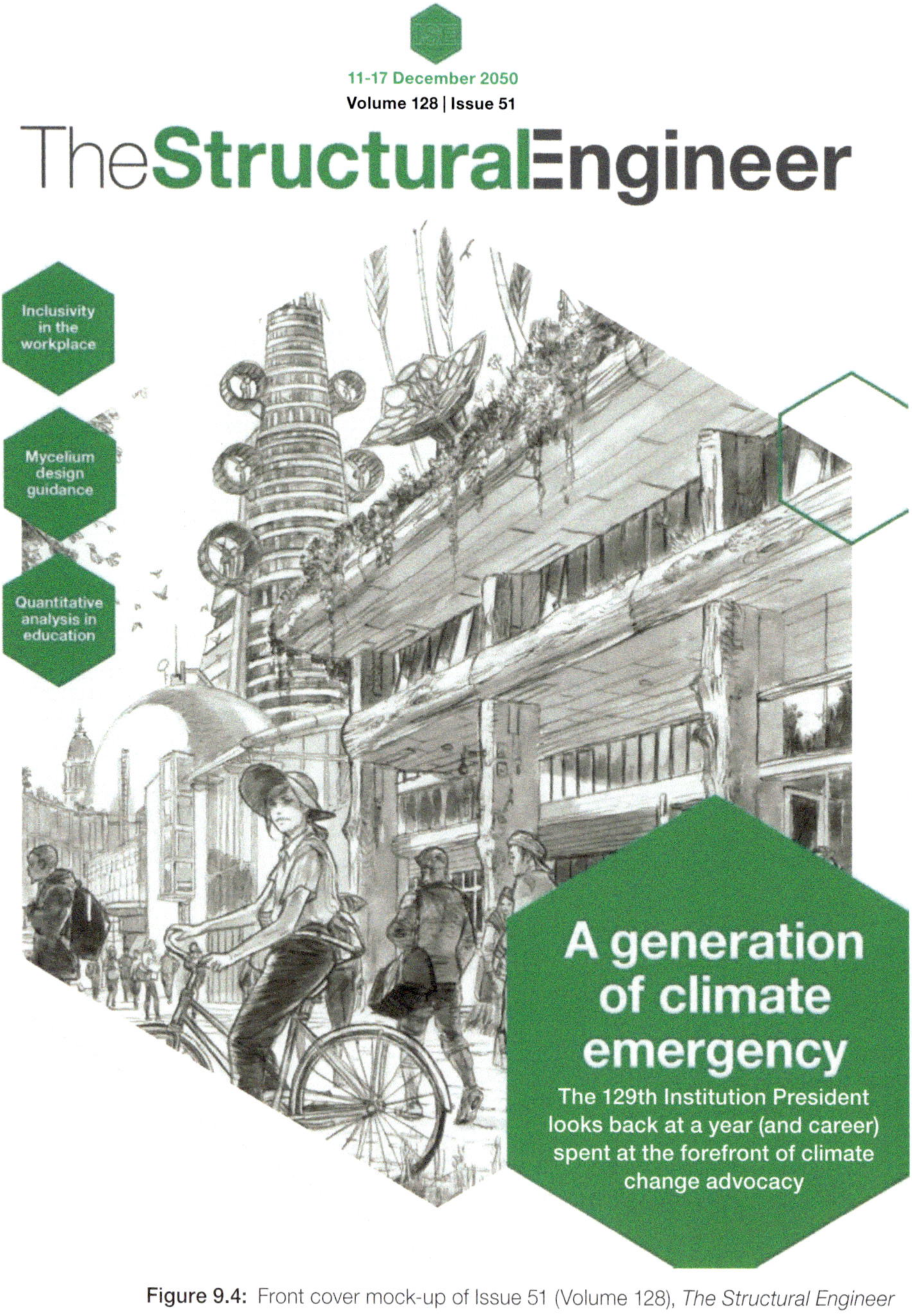

Figure 9.4: Front cover mock-up of Issue 51 (Volume 128), *The Structural Engineer*

Contents

Upfront

Global challenges

Professional guidance

Technical

System focus

Opinion

At the back

Figure 9.5: Mock-up of contents page of Issue 51 (Volume 128), *The Structural Engineer*

Notes on the contents

Author names:

The authors attributed to these articles are fictional — but are included to represent the increasing diversity of the profession in terms of race and gender, and a move towards capturing a more global understanding of changes in the industry.

Upfront/At the back:

- While we see much of the magazine's format remaining the same (and we are especially keen to see the celebration of incredible drawings retained) there are likely to be subtle shifts of focus during the next 25–30 years

Global challenges:

- In 2050 there will still be many global challenges — not just the climate emergency — the change in title reflects this
- Although there is investment in digital infrastructure, degradation of data is still not fully understood. The futurist, Douglas Coupland, has written about this in a collection of essays[9.10]

Professional guidance:

- This is a look back on the intervening 30 years since the article *The future is human*[2.11] was published in *The Structural Engineer* in 2020

Technical:

- Mycelium is just emerging as a material. Current work is in boarding materials (and fashion) but in 30 years' time its application may be much broader[9.12]

- Despite being highly susceptible to earthquakes, earth structures are used widely around the world — and may become more utilised in the future. Increasing resilience of these types of building will be of growing importance
- There is something beautiful and joyful about doubly-curved timber — and I hope the future will also include work that delights in this way

Several of the current sections are likely to change:

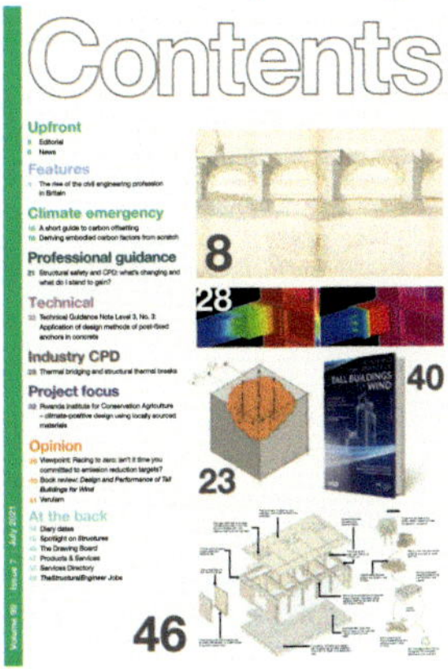

Contents

Upfront
Editorial
News

Features
The rise of the civil engineering profession in Britain

Climate emergency
A short guide to carbon offsetting
Deriving embodied carbon factors from scratch

Professional guidance
21 Structural safety and CPD: what's changing and what do I stand to gain?

Technical
Technical Guidance Note Level 3, No. 3: Application of design methods of post-fixed anchors in concrete

Industry CPD
Thermal bridging and structural thermal breaks

Project focus
32 Rwanda Institute for Conservation Agriculture – climate-positive design using locally sourced materials

Opinion
Viewpoint: Racing to zero: isn't it time you committed to emission reduction targets?
Book review: *Design and Performance of Tall Buildings for Wind*
Verulam

At the back
Diary dates
Spotlight on Structures
The Drawing Board
Products & Services
Services Directory
TheStructuralEngineer Jobs

Volume 99 | Issue 7 | July 2021

8 28 40 23 46

Systems focus:

- In 2050, systems thinking will be an integral part of all design thinking — and as a result we will move from 'project focus' to 'systems focus'

Opinion:

- We are sure that even by 2050 there will still be much for designers to do in terms of working towards regenerative outcomes — with many thoughtful and provocative conversations to be had
- However, we also believe that many topics of conversation will be similer to those held today, and even to those held 25–30 years ago (indeed, we hope that one such topic will be the education of future engineers)

9.4 ON SITE AT A SCHOOL

Welcome to David Attenborough High School. Today we will be doing some structural integrity testing of an outdoor classroom that the students have built themselves. We have a long history with the school. We have done a variety of reuse projects and have carried out extensive checking of the existing structures to extend the design life and assess revised loadings. We now have a detailed database of information which makes future assessments straight forward. We can use probabilistic methods to check the capacity of the existing structure, which enable us to make much less conservative decisions.

We have reviewed changes in loading for increased water storage, but also for changes in the curriculum. Schooling now is much more practical. Students have to build their own keyboard before they can learn to type. And as indoor growing becomes more common, especially in areas of high population density, all schools now have growing labs where students learn to grow and monitor their own food production[9.13]. For some, it gives them access to food that they don't have at home, and we are trying to work out how we can improve access over the holidays.

Food costs will escalate reflecting the effort required to produce it. As a result self-production will be essential. Food poverty will still be an issue.

We have enabled the addition of extra stories on the roof. We had to carry out detailed testing to ensure the existing roof slab had capacity. We used AI to carry out a series of probabilistic analysis using quantum computing and parallel processing[9.14]. But in honesty, we only do this when necessary, most of the time a good sketch and $wl^2/8$ is enough.

For the new storeys we try and design with lightweight, renewable and flexible materials. The reduced weight obviously helps to keep the loading down. The renewable materials are part of the local sourcing strategy. We could use abundant materials like earth, but not on top of an existing building. We have carried out some trials with bamboo, but for now timber remains our preference due to local availability of materials. We also now try to make structures flexible. Rather than design for limited deflection and vibration we found that we can achieve much more materially-efficient solutions by allowing structures to move. It took a while to resolve some of the waterproofing and façade details, but the carbon savings have been significant. Rigid structures are now limited to buildings of future historical importance and a permit is required to design in that way. We can still use steel and we are starting to see concrete slowly return after it was banned in the 2030s. But these materials are used sparingly and are saved predominantly for rigid structures.

We use blocks created from aggregate that is a product of carbon-capture technology and cement-free binders to lift the new storey out of the wet zone to prevent rotting. In addition, all the materials are grown locally and, where possible, fixings are also locally-grown timber nails.

We also monitor the structure. We have highlighted areas of risk and have attached strain gauges and accelerometers which alert us to any significant displacement or change

As we move to steel made from renewables only (due to changes in legislation and a move to full circularity), its cost will increase — resulting in a fall in demand. We will continue to see steel manufactured but in smaller quantities as we look to reuse as much as possible and also look for local alternatives.

While the concrete roadmap is encouraging, if we are really going to achieve absolute zero we can't (according to some) keep making concrete. A move away from cement towards waste materials will be important — and I don't see concrete ever disappearing — but without incentives or bans our ability to innovate is very limited, as the last 40 years have shown us.

The use of carbon capture is hotly debated. In an optimistic future it seems reasonable to consider carbon capture directly from source. The potential for more general carbon capture seems less certain. If we can capture carbon we could then consider turning it into useable products, and hence 'build with carbon'.

In *Designing Regenerative Cultures*, Wahl refers to not being overreliant on digital infrastructure and building resilience through analogue systems.

in stiffness. The pupils are also required to take analogue readings to ensure the electronic ones are working, as most are a decade or more beyond their working life. This approach helps us to monitor the health of the building and ensure we don't overload the structure. It also alerts us to long-term movements and changes in soil conditions due to climatic impacts.

Today's visit though is a little more fun. Every student is involved in designing and building their own classroom. We carried out an initial assessment to ensure the design didn't pose any major risks. And now we are back to inspect the build before they utilise it.

This idea is based on a series of projects that Integral Engineering used to do with the WOMAD festival, where people would come early and build structures which were then used as part of the festival (Figure 9.6). I was involved one year as a checking engineer.

The building is constructed from waste and reclaimed materials. Each year's project can utilise materials from the classroom which was constructed the year before, and the first job is to deconstruct and inventory what is available. The students then source other waste that can be used in the construction — anything that isn't used must find a new place in the ecosystem, nothing can be wasted. It is very challenging, but highly rewarding. People use leftover materials to make furniture, clothes, sculptures. Students are involved across a wide number of disciplines, but everyone participates, and the buzz around the project is incredible.

This is a different type of engineering again. We are not able to carry out lots of detailed

Figure 9.6: Tent at WOMAD festival

analysis, there is no probabilistic design. Instead, we use our intuition and experience, and at 72 I sure have a lot of that! We load-test the structure by pushing it and pulling it. We have some simple tools, like instrumented ratchet straps that provide force displacement readings. The main aim is to consider how ductile failures will be, what warnings will the structure provide (and we can add some electronic tags to help with this) and where the risks lie? We normally limit use to windspeeds below 40 km/h to be safe.

Whenever I visit the school there is always another task that needs to be carried out. I try and distribute the work fairly across the different engineering cooperatives in the city. We all share our work, it helps to build robust relationships, I'm not going to live forever and I certainly hope to retire at some point soon, at which point a new main contact will need to step forward and take over the role. The aim isn't to use the relationship to maximise profit, but to build long-term and sustainable workflows. By sharing projects, we can work together in search of abundance. For a while people were worried that the role of the engineer would disappear with the advent of AI and enhanced computing, but in reality, it has just led to a diversity of jobs, more creativity, more hands-on experience — and the role of an engineer now much more closely matches how I imagined it would be when I was a boy; and that's great.

9.5 EXISTING BUILDING EXTENSION

As we head home, I'd like to take us on a short detour to visit a recent project I worked on. The project took a series of existing buildings and extended them upwards, utilising the additional capacity of the structure and the foundations. As the building was 25 years old, much of the settlement (for which the foundations had been designed) had already occurred. We converted the basement, which was a car park, into a mixture of plant rooms, which were located on slabs that were directly bearing on rock, so had lots of spare capacity. This created more capacity in the structure above, as previously the plant rooms had been located on the roof. We also used basement levels as growing space for urban farming. There are some emergency cool rooms for the vulnerable, who find it hard to shelter from the heat in the peak of summer now. The ground floor remains as retail but mixed with workshops and living spaces for birds and insects. New 'green corridors' have been created enabling wildlife to move freely around the city, including some that pass through the buildings. The workshops carry out a variety of functions, from fixing clothes and mechanical devices, to teaching people new skills. The floors above are now office space, with a new two storey residential development on top of the existing building. The offices are only inhabitable 95% of the year due to the heat, but rather than spend significant sums on retrofitting better cooling systems (and the associated energy cost) there is a more flexible working culture now, so people don't have to work when it's too hot.

This idea emerged from a Buro Happold-hosted workshop on 'the future of design'.

The residential upward extension is constructed to be as light as possible. Timber platform frames have been used which both spread the load and are very lightweight. Where longer spans are necessary, CLT panels with glulam stiffening ribs have been utilised, but these are made from reclaimed timber which is locally upcycled. There are a few very large spans, made with reclaimed steel. The building's non loadbearing walls are made from a mixture of natural building materials, including hemp shiv and mycelium-derived products. The locations of these materials are optimised for loading, acoustic and thermal performance, as well as their different aesthetic requirements.

The main challenge, as you can imagine, was carrying out all the testing and analysis of the existing structure to prove it was adequate for the additional load. But as reuse has become the norm, the guidance and expertise are now more readily available than for new build. 'Heat maps' show where significant loads can be accommodated and where they should be avoided. Deflection and vibration analysis are carried out, but not so we can stiffen slabs, but instead optimise layouts to work with the building response (you don't want your bed where the floor moves most as you feel vibration much more when lying down) (Figure 9.7).

Figure 9.7: Vibration testing of the existing slab (left) and strength testing of a concrete sample (right)

Working on these types of projects is incredibly satisfying. It brings many of my engineering skills to bear. Seeking opportunities. Working with communities. Enabling positive outcomes. Specifying local materials. Physically testing materials. Carrying out complex analysis. Communicating solutions to technical and non-technical stakeholders. Of course, no one person can do all of this, which is why working collaboratively is both so important and so enjoyable. In fact, for me, if I had to highlight one thing I love about the job, it is the diverse teams that I now work with, and the friendships I've made over the years. If I was being honest, I would say I was proud to have worked on this (and many other) projects. I am proud of what we have achieved together.

REFERENCES

9.1 Figueres C and Rivett-Carnac T. *The Future We Choose: Surviving the Climate Crisis*. London: Manilla Press, 2020.

9.2 Hopkins R. *From What Is to What If: Unleashing the Power of Imagination to Create the Future We Want*. White River Junction (Vermont): Chelsea Green Publishing, 2019.

9.3 Chatterton C. *Unlocking Sustainable Cities*. London: Pluto Press, 2018.

9.4 Wahl DC. *Designing Regenerative Cultures*. Bridport: Triarchy Press, 2016.

9.5 The Institution of Structural Engineers. *Design for zero*. London: IStructE Ltd, 2021.

9.6 The Institution of Structural Engineers. *Circular economy and reuse: guidance for designers*. London: IStructE Ltd, 2023.

9.7 Minimising Energy in Construction (MEICON). *Survey of Structural Engineering Practice*. Available at: www.meicon.net/survey2018 [Accessed: December 2023].

9.8 The Institution of Structural Engineers. *Inclusive design for structural engineers*. London: IStructE Ltd, 2023. (in production)

9.9 BBC Earth. *The Green Planet*. Available at: www.bbcearth.com/shows/the-green-planet [Accessed: December 2023].

9.10 Coupland D. *Bit Rot: short stories and essays*. London: Windmill Books, 2017.

9.11 de Hoog T. 'The future is human.' *The Structural Engineer*. 2020:98(1). Available at: https://doi.org/10.56330/HTMI1094 [Accessed: December 2023].

9.12 Dezeen. *Blast Studio 3D prints column from mycelium to make "architecture that could feed people"*. Available at: www.dezeen.com/2022/01/18/blast-studio-tree-column-mycelium- design/ [Accessed: December 2023].

9.13 Superflux. *A pragmatic experiment practicing hope for a future disrupted by climate change*. Available at: https://superflux.in/index.php/work/mitigation-of-shock/# [Accessed: December 2023].

9.14 The Institution of Structural Engineers. *Computational engineering*. London: IStructE Ltd, 2021.

A note of thanks

Many of the ideas in this book have been developed through conversations with the first cohort of the Constructivist Regenerative Design Lab. To Laura Batty, Emma Crichton, Nick Francis, Milly Last, Tim Lucas, Eva Macnamara, Cíaran Malik, Martyn Mclaggan, Ellie Osborne, Si-Joe Portal, Jessica Rowe, Mike Sefton, Suzannah Sherman, Gemma Small, Alexie Sommer, Olivia Sweeney, Karen Winfield, Chris Wise we owe a debt of thanks.

Thank you to Buro Happold who let us present a very early version of this book and gave us lots to think about in the conversations that occurred as a result. Thank you to all the other companies whose work has featured in the book. Being a regenerative practitioner is not an individual pursuit but depends on community. If you or your work is in this book it is because we respect and acknowledge the work that you do and the impact you are having on our community.

We are also extremely grateful to Will Arnold, Kai Chelliah, Mike Cook, Emma Crichton, Nick Ely, Patrick Hayes, Peter Ho, Phil Isaac, Niamh McCloskey and Chris Wise who, as a result of reading multiple draft versions between them, gave us much encouragement (and so many great suggestions for improvement) along the way.

Oliver would also like to thank:

The Royal Commission for the Exhibition of 1851 who, by appointing me 1851 Fellow in Regenerative Design, have given me the confidence and means to write this book. Thank you also, to the Sir Misha Black Awards Committee and Engineers Without Borders UK who supported the first Lab.

Huge thanks to my Constructivist colleagues, Ellie Osborne and Alexie Sommer with whom I have set up and run the Lab. To my mentor Rowan Conway for guiding me on this journey. To Chris Wise for 15 years of championing me. To Will Arnold for his tireless curiosity. To Nick Francis for his clarity of thought and willingness to talk 'systems' at any time of day. To Joel de Mowbray and Morgan da Silva at Yes Make for showing us how all this is done. To Michael Pawlyn for his support of our work. To Bill Sharpe for our enlightening conversations. And to Hazel Will Wood for being a place of inspiration.

And finally I am indebted to my partner Mary Stevens, who is consistently two years ahead of me in her ideas. I am in awe of your thinking, grateful for your intellectual and emotional support and for your patience while I catch up.

James would also like to thank:

All my colleagues at Ramboll (previously Whitbybird) where I spent 12 years of my life practising as an engineer. I learnt a huge amount over this time. While I can't include everyone, I would like to especially thank David Burns who mentored me for many years, Nicholas Porter, who first challenged me to think about sustainability, and Ellen Grist who took it to another level. Thank you to all my colleagues at Integral Engineering Design — you have all had a profound impact on me and I hope you can see it reflected in this book. Thank you to my current employer, the University of Bristol, and my head of School, Simon Neild — it has only been possible to find the time to write four books in eight years because you put trust in me and gave me the space I needed to think and write. Also thank you to my colleagues in BILT (Bristol Institute of Learning and Teaching) who published my blogs as I wrote this book.

Thank you to everyone who aided me in the writing — particularly to James Solly, James Kitchin, Kelly Alvarez Doran, Luis Felipe Lopez, Kent Harries, Dave Taylor, Joseph McFarlane, Paul Jaquin, Ed Metters, Rosie Goldrick, Craig White and Jez Ralph who made time to share their work and wisdom with me — the book is much richer and more global for your collective input. Thank you to Kelly Harrison and Simon Smith who both presented to my students and who I subsequently acquired ideas from (and thank you Simon for approaching me to write my first book — it gave me the confidence to go on and write more).

Finally, thank you to my wife and family. Regenerative design is another way to talk about kindness. A huge debt of thanks goes to my wife Lizzie who constantly astonishes me by demonstrating and enabling kindness in the community around us.

Part 4
Appendices

APPENDIX A

The 12 Levers of Systems Change

In the writing of the book we found it helpful to utilise the work of Donella Meadows, and specifically the 12 Levers of Systems Change[A.1].

We went through the 12 levers and tried to think of examples in our industry. The following is not an all-encompassing list, but hopefully illustrates the different levers in a way that is applicable to an engineer.

Meadows lists the levers in reverse order, starting with the least effective for change and working towards the most effective. We have followed the same logic and numbering system here:

12. Numbers

As engineers we have a strong affinity with numbers. However, counting things does not bring about significant change. Here are three different measures we use in our profession, and why, on their own, they don't achieve significant change:

BREEAM credits: The act of trying to accumulate credits for the desired BREEAM rating. If the decisions made are predicated on achieving the required number of credits and are not embedded in a greater purpose, then accumulating BREEAM credits is no guarantee of change. However, see levers 8, 7 and 3 for examples of BREEAM effecting change.

Safety factors: Once we know a building is safe, we may consider how efficient (or inefficient) a building is, but as MEICON[A.2] shows, the reality is that engineers are happy to increase the capacity by a further 100%.

Carbon counting: Trying to calculate the embodied carbon on a design, after the initial concept design has taken place and the design is mostly set — therefore removing most of the opportunity to make meaningful reductions.

11. Buffers

Systems are dynamic. In structural dynamics, damping changes how quickly a system responds to change. Likewise in systems. For example:

Availability of materials: Buffers mean that our building materials are readily available — but they also impede innovation. For example, if I want to use a new product, what is the lead-in time? Can I be 100% confident the material will be available when I need it?

Project planning: Buffers mean we can be flexible about work, allowing us to pick up other issues (e.g., urgent onsite queries). If we don't have enough buffers the quality of work may drop, if the buffers are too big we may spend more time and money on a project than necessary.

10. Stock and flow structures

In systems, the numbers we count and the size of buffers matter because of stocks and flows. For example, in reuse of building materials currently, the level of reuse is limited because we have no way to store and transport materials (low stock therefore low buffer). We rely on a project needing them at the precise time we deconstruct, otherwise we consider other forms of waste management (e.g., downcycling to rubble, recycling into steel or burning and capturing the energy). But if we had the infrastructure to store and transport materials we could change this.

9. Delays

Delays in the system can lead to two problems. If they are too slow they can lead to points where irreversible damage can occur. Take carbon emissions and the climate emergency for example; we have delayed action while we ensure we trust the data, and now the chances of reacting in time to limit warming to 1.5°C are rapidly shrinking towards zero.

If actions are too quick, they can lead to overreaction and waste. If you are collaborating on a BIM model and the architect makes a change, this could trigger a flurry of activity as you quickly resize beams and columns, only for them to move it back the next day (or hour or minute).

8. Balancing feedback loops

A balancing feedback loop requires a goal, a method of monitoring and signalling and a response mechanism. The classic example is a thermostat in a house.

By setting BREEAM as a planning requirement it sets a balancing feedback loop. While counting BREEAM points has little impact, having a goal (the level of BREEAM required), monitoring and signalling (the record of credits, which includes a risk assessment to manage different credits) and establishing a response mechanism (often credits you can 'buy' to achieve the required goal if other credits cannot be acquired) is certainly impactful. The difference here is that this isn't about counting — it is the desire to achieve a goal.

Likewise setting an embodied carbon target. If we carbon count with a goal (set by planners or others) using the SCORS system then the aim is not just to count, but to do so with a purpose — and the purpose should drive the design. The stronger the goal, the harder it is to deviate. For example, to achieve the goal we may reduce column grids, revise loading requirements (still within code, but without additional factors applied), design with greater efficiency or change materials. By adding a balancing feedback loop, we can start to see a greater impact.

7. Reinforcing feedback loops

Reinforcing loops are what we are talking about when we refer to positive and negative cycles.

Taking the example of carbon counting or BREEAM, a reinforcing loop might look like a number of large consultants all achieving a higher-than-expected standard and then telling everyone about it.

By declaring their SCORS rating or BREEAM rating, this challenges others to do the same. As more people declare their project rating, the reinforcing loop makes it harder and harder for others not to join in. This may trigger the codification of the approach. For example, BREEAM has not always been a planning requirement. Originally it was something that projects opted to do, but as more projects declared they were very good, or excellent, or outstanding, the more it was deemed necessary for projects to have a rating (and a good one) which then enabled planners to start insisting on them.

6. Information flows

Have you been to a building site and seen a sign declaring how many days it's been since the last accident? This is a form of information flow. By declaring this information at the entrance to the site the contractor is doing multiple things: they are alerting you to the fact that this is a dangerous environment; they are alerting you to the fact they record and account for accidents, and they are showing that they want to reduce the number of accidents (the more days since the last accident the better). This information flow changes behaviour on site. Likewise, once you have worked on a site where they are proud of their health and safety record, why would you want to work on one where health and safety is not valued? So this information flows from site to site. Of course, accidents do happen, but we can all learn from them by sharing the details (confidentially) with CROSS (Collaborative Reporting for Safer Structures)[A.3].

Likewise, as we inform people about the embodied carbon on a project we enable people to start calculating the impact of their projects. So, the information in The Inventory of Carbon and Energy (ICE)[A.4] and *How to calculate embodied carbon*[A.5] guides are invaluable. For a long time, this information was locked up within specialist or individual practices, but now it is available to all.

As a counter example, engineers generally appear more concerned and active in the climate emergency than they are in the biodiversity emergency. As we discuss in this book, we could consider embodied biodiversity — and by making this information known we may discover that we have greater impact than we currently imagine. Attempts to share information on embodied biodiversity are starting to emerge[A.6, A.7].

5. Rules

Rules are one of the easiest levers for us to understand and apply, as our industry has lots of them. Codes of practice, building regulations and health & safety rules all define what we do. We could reduce carbon by designing to a factor of safety below that permitted by the rules, but we don't, because it wouldn't be safe and the rules do not permit it.

Similar rules could exist for carbon reductions, biodiversity gain and design for reuse.

4. Self-organisation

In response to the global advocacy that emerged in 2019, the IStructE started to self-organise around the climate emergency in early 2020. Central to this was the creation of the Climate Emergency Task Group, which has gone on to produce and collate a wealth of invaluable resources[A.8]. To enable this to happen, far from setting strict rules, you need to enable the environment for serendipity and diversity. To allow self-organisation means letting go of control!

3. Goals

As noted in the 8th lever, the goals we set define how we work, the decisions we make and the outcomes we try to achieve. BREEAM was set up off the back of a project, BedZED[A.9], which had a number of interconnecting regenerative goals. Does this make every project with a BREEAM rating regenerative? No. Because in converting the goals into rules, feedback loops and numbers many of the goals have been lost. By establishing and working towards those goals we bring about the biggest change.

As engineers we have a primary goal — to make our buildings safe, closely followed by goals around the project being completed on time and within budget. But what if we set the goal from Chapter 3 — **for human and living systems to survive, thrive and co-evolve**? We would still have safety, but we would also think very differently about the work of the designer.

2. Paradigms

We live in the infinite growth paradigm. Resources are seen as not finite but infinite. If it looks like we are running out of something, we just need to dig deeper, look further afield. But despite this being the paradigm of our age, it is not the case. We have abundance and renewal, but at a set rate. We live on a bounded planet, which has a very large, but still limited, energy input from the sun. For billions of years that energy has been stored up. Converted into all sorts of things. But now we are using it, and at a much quicker rate than it was created at. We are beyond 'one world living'. The impact of this is becoming well documented, and we now need a new paradigm. This has been presented in many ways, but the doughnut economics diagram (Chapter 2), of living within the Earth's resources while also raising social standards for everyone, is a clear way of thinking about it. The UN's Sustainable Development Goals[A.10] are also a good reference point.

We also live in the scientific paradigm. This paradigm has led to great advances in human prosperity, but it is not without its downsides. And other paradigms exist e.g., those of different Indigenous and religious groups.

1. Transcending paradigm

The ability to hold multiple paradigms simultaneously and see them all as different ways of knowing. This space is harder to conceive and describe. However, in *Braiding Sweetgrass*[A.11] the author begins to wrestle with this space.

REFERENCES

A.1 Meadows DH. *Thinking in Systems.* London: Chelsea Green Publishing, 2015.

A.2 Minimising Energy in Construction (MEICON). *What we do.* Available at: www.meicon.net/what-we-do [Accessed: December 2023].

A.3 CROSS UK. *Share knowledge to help create a safer built environment.* Available at: www.cross-safety.org/uk [Accessed: December 2023].

A.4 Hammond G and Jones C. *Embodied Carbon: The Inventory of Carbon and Energy (ICE).* 2011. Available at: greenbuildingencyclopaedia/wp-content/uploads/2014/07/Full-BSRIA-ICE-guide.pdf [Accessed: December 2023].
A.5 Gibbons O and Orr J. *How to calculate embodied carbon.* London: IStructE Ltd, 2020.
A.6 Expedition Engineering. *The Embodied Biodiversity Impacts of Construction Materials.* Available at: https://expedition.uk.com/project/embodied-biodiversity-impacts-of-construction-materials/ [Accessed: December 2023].
A.7 UKGBC. *Embodied Ecological Impacts.* Available at: https://ukgbc.org/our-work/topics/embodied-ecological-impacts/ [Accessed: December 2023].
A.8 The Institution of Structural Engineers. *Climate emergency: Sustainability Resource Map.* Available at: https://www.istructe.org/resources/climate-emergency/ [Accessed: December 2023].
A.9 Bioregional. *BedZED — the UK's first major sustainable community.* Available at: www.bioregional.com/projects-and-services/case-studies/bedzed-the-uks-first-large-scale-eco-village [Accessed: December 2023].
A.10 United Nations. *Sustainable Development Goals.* Available at: https://sdgs.un.org/goals [Accessed: December 2023].
A.11 Kimmerer RW. *Braiding Sweetgrass: Indigenous Wisdom, Scientific Knowledge and the Teachings of Plants.* London: Penguin, 2020.

APPENDIX B

Definitions of 'regenerative design'

As highlighted in Appendix A, setting a goal is incredibly powerful in terms of leveraging change — which is why we opted to present a regenerative goal rather than a definition in this book. However, we are also aware that some people may request a definition of regenerative design. Here are three that we have found helpful:

1. RESTORE define 'regenerative design' as:

   "Enabling social and ecological systems to maintain a healthy state and evolve".[B.1]

This one sentence definition is, for us, the simplest formulation that captures the key characteristics of regenerative design:

- **Enabling** — emphasises giving capacity to rather than doing it yourself
- **Social and ecological** — emphasises the connection between human systems and wider living systems
- **Maintain a healthy state** — not just restored to a healthy state, but enabling that system to stay in a healthy state thereafter

- **And evolve** — also enabling the system to adapt as conditions change, thereby restoring system resilience

They go on to say:

> "It is not the building that is 'regenerated' in the same sense as the self-healing and self-organizing attributes of a living system, but by the ways that the act of building can be a catalyst for positive change within the unique 'place' in which it is situated".

So, it isn't the building that is 'regenerative' but rather the process of designing, building and utilising it.

2. Mang and Reed (from the Regenesis group) describe 'regenerative design' as:

   "A system of technologies and strategies, based on an understanding of the inner working of ecosystems that generates designs that regenerate socio-ecological wholes (i.e., generate anew their inherent capacity for vitality, viability and evolution) rather than deplete their underlying life support systems and resources.[B.2]

This definition echoes many of the characteristics in the RESTORE definition, and brings in two additional ideas:

- It is based on the understanding of ecosystems — emphasising the importance of observation and connection to nature
- The ability for human-natural systems to maintain a healthy state and evolve described as 'socio-ecological wholeness'

This image of wholeness can be helpful for visualising how a system might be incomplete and what might be needed to make it whole again.

3. The third description of 'regenerative design' comes from Carol Sanford:

   "Whereas regenerative design begins from the unique and specific potential of a situation and works to develop the capacity to bring this potential forward, design thinking emphasises problem-solving as a way of arresting disorder. In other words, the two operate at different levels of paradigm".[B.3]

We have included this description in Chapter 3. It emphasises that regenerative design is quite different from how we might traditionally design — here we are working at a higher level of the system, at the level of paradigm change.

REFERENCES

B.1 RESTORE. *Sustainability: Restorative to Regenerative.* Available at: https://fairsnape.files.wordpress.com/2021/02/sustainability-restorative-to-regenerative.pdf [Accessed: December 2023].

B.2 Mang P and Reed B. *Regenerative Development and Design.* [In: Meyers, R.A. (ed) *Encyclopedia of Sustainability Science and Technology*]. New York NY: Springer, 2012. Available at: https://doi.org/10.1007/978-1-4419-0851-3_303 [Accessed: December 2023].

B.3 Sanford C. *The Regenerative Life.* Boston MA: Nicholas Brealey Publishing, 2020.

APPENDIX C

Creating your own artefact from the future

Flourish[C.1] provides a whole chapter on 'possibilism'; the act of imagining the future and then working towards that future. Or, as we have referred to it in this book 'Horizon Three (H3) thinking'. For some it is hard to imagine H3 and they are happier dwelling on H2. For others it is the other way around, the future may be clear, how we get there is the challenge. Either way we need some idea of where we are going.

In this appendix we outline the approach we took in creating the epilogue (Chapter 9) for this book. Our 'artefact from the future' started life as a 'future walk' — and included not just the final locations, but also details of the walk between these places.

Setting a future date

2050 is a key year for all of us. It is the year that we have set to get to zero. The year is, of course, a little arbitrary, but it is far enough into the future that it is hard to know exactly what it will look like, yet close enough in the future that it will be in many of our lifetimes.

We don't know what 2050 will look like, but we do know it will be very different to now. If we choose to act quickly and positively on climate change (and the other areas where we have overstepped our planetary boundaries, such as biodiversity loss) we know that things will be different by choice. If we choose not to act, they will also be different by choice, but in a different (and bleak) way. *The Future We Choose*[C.1] paints vivid images of what 2050 'best case' and 'worst case' scenarios might look like[C.2, C.3].

Future walks

A future walk imagines a specific place in 2050. It encourages us to either physically or imaginatively move slowly around that place.

Figure C1 is James' street in the early/mid 2020s. He has lived in this street for more than 20 years. He knows a lot about the lives of the people, the construction of the houses, the changing shape of the building fabric and the ecosystem around him.

Figure C1: James' street today

Figure C2 is how he imagines that place will look in 2050. At this point the details are unimportant. The key issue is that this future imagining comes from a deep understanding of a place developed over time, through observation and being. He understands some of the socio-ecosystem.

Figure C2: James' street in 2050

As we imagine 2050 if we choose to design a world which works with our socio-ecosystem and looks to create the flourishing of both people and nature in a way that goes from doing less harm to doing good, having some understanding of our socio-ecosystem is important.

So, we walk (the act of walking sets us in a place) in a place we know and imagine how it might be in 2050.

The act of walking also enables us to visit places of importance. As a structural engineer these places are building sites, my place of work, my home (community) and the spaces that connect these, where we can move through the landscape and see the different changes.

Finally, the act of walking means going slowly. Not rushing, taking your time. Pausing to look. Sitting and dreaming for a moment.

Of course, we don't have to walk (and for some people it isn't possible), the idea is simply to move slowly through the landscape.

Is it hope or fantasy?

One of the purposes of a future walk is to give people (including ourselves) hope. Hope is an important creative ingredient. If one loses hope it becomes hard to do the work needed to bring about change. If we lose hope, then we can become overwhelmed by anxiety. We have experienced both. But we remain hopeful.

However, while hope is helpful, fantasy is unhelpful. It is a fantasy to assume that we can continue to operate exactly as we are, and the world will continue to be as it is. We cannot continue to emit greenhouse gases and expect that the atmosphere won't change (with all the impacts that leads to). We cannot continue to damage or remove habitats and expect biodiversity to stay constant.

Everyone will have their own view on how optimistic or realistic we should be when imagining the future. Personally, we think it is unhelpful to assume that our lack of action over the last 40 years will have no consequences. When we imagine the future, we have therefore tried to include some of these consequences. We know, from receiving feedback from people who have seen our ideas, that they find the consequences uncomfortable — and sometimes terrifying. We hope that these consequences won't materialise, but we think it is useful to have discussions. The one that is most troubling is a move away from a libertarian approach to governance and towards a draconian approach. Our own thinking on this has been greatly impacted by the COVID-19 pandemic and particularly the lockdowns. However rather than paint a picture where that is the end outcome, much like the lockdowns, our narrative suggests that in the transition some of our liberties may be removed, for a time. This is not an ideology that we support (on the political compass James is strongly libertarian and left wing) but one that might come to pass as a consequence of our current inaction.

The other thing to consider is your hopefulness around technology? Are you a techno-optimist that believes technology will save us? Will carbon capture (or another technology) solve all our problems? Again, we received a lot of feedback on this — and again we tried to set a middle ground — in 2050 there will be some carbon capture, but it will have come late and it will be utilised in discreet instances (for example to capture carbon generated from turning human waste into fuel). You may have a very different view.

Or, as is the case in *The Future We Choose*[C.1], you might want to create different cases; maybe a best case/worst case and a realistic/optimistic case.

How to create a future walk

So how do you create a future walk?

Step 1: Immerse yourself in ideas.

When creating our future walk the first thing we did was immerse ourselves in other people's ideas. We read other people's visions on the future. We read other people's ideas on regenerative design, and circular economies, and climate activism (see Appendix D for a list of the most useful and relevant books). We read widely and tried to capture ideas. We read magazines from diverse fields. We looked at art projects. Listened to podcasts. Watched videos of people's vision of the future. And then we read some more. We also shared our ideas with each other. We pushed them backwards and forwards.

This includes a subscription to Stack[C.5] which provides a different magazine every month. The magazines are very diverse, from across the globe, and cover a huge variety of topics. They are unified by their commitment to graphic design. The magazines are often challenging, which helps understand a more diverse set of global views.

Step 2: Pick a location and walk around it.

James immersed himself in the socio-economic location where the future would be based. He chose the neighbourhood where he lives. For work he commutes by train which involves a walk to and from the station every day. When he works from home, he walks anyway to get some exercise and observe a place. The act of walking creates time to observe and think. Cycling or driving would give a different impression. You can do this both physically and imaginatively. We have done both and found them equally rewarding.

Step 3: Ask how the ideas of the future might work in this location.

Having picked a place and started to research, you can start to dream what this place might look like in 2050. We love dreaming as we walk from place to place — so this works for us, but it may not for you. But find a way to put place and ideas together.

Step 4: Draw/write/narrate/discuss your thoughts.

Narrate a space (you can pretend you're chatting to someone on the phone) or draw it and make loads of notes, or jot down notes as you walk. Or walk with someone else (or sit with someone else and imagine together) and discuss ideas. Whatever helps you to capture your thoughts about a place in 2050. Maybe take photos of the place now and add annotations (Figure C3).

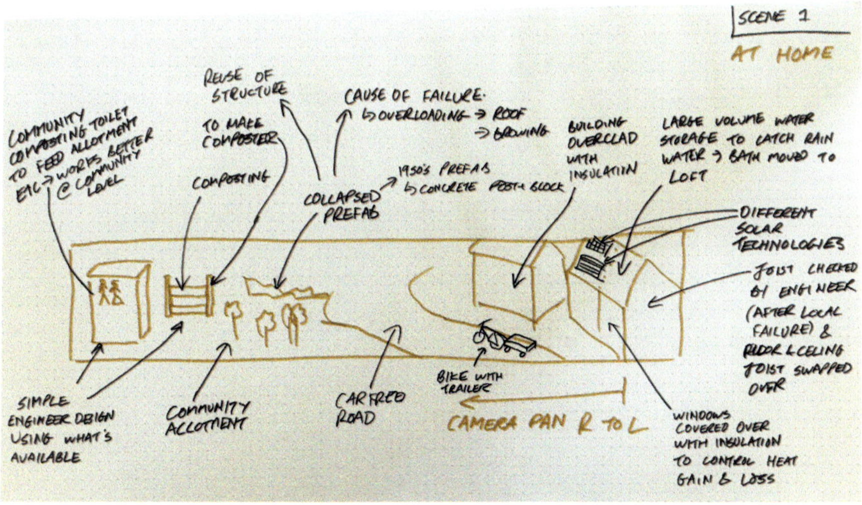

Figure C3: Initial sketch capturing the aspects that might be present on James' street in 2050

Step 5: Share your ideas.

Finally, we found it helpful to think in a series of scenes. You don't have to capture everything in one place. Instead, imagine standing somewhere explaining to someone what the world might look like in 2050. Consider writing down a narration. Add photos or storyboards.

As engineers it is useful to think not just about the big picture but the details. We don't have to get all the details right, but the more detailed the future we can map, the more helpful the picture is to others.

Once done, share it with others. As a lecture, or a video, or a podcast, but don't keep it to yourself.

At the 2022 IStructE annual academics' conference, James shared his own vision of the future, but also worked with student ambassador of Engineers Without Borders, Cleo Parker, who presented her own vision. Cleo described the hope of meeting ambitious global targets, like the UN Sustainable Development Goals[C.6], and the pride of working and shaping integrity-driven companies; where daily choices were determined by delivering on their international responsibility. In these companies, inclusive practices were everyday norms and attracted many more people to work in engineering. Participatory methods were fundamental to how projects were run. You can listen to both views in a presentation available from the IStructE website[C.7].

What to do if you're stuck

One strategy we have found helpful when stuck is to imagine a different context. For example, what if we lived on Mars and had no outside supply? Everything had to be reused, recycled and upcycled. The only inputs we have are the energy from the sun and the regolith on which our settlement is built. Now imagine this is on Earth — what could we add to increase both human and natural flourishing? What would we choose to not add? This scenario was vividly brought to life in a public art project in 2022[C.8]. By changing the context we can think about the problem in a different way.

REFERENCES

C.1 Ichioka S and Pawlyn M. *Flourish: Design Paradigms for Our Planetary Emergency.* Bridport: Triarchy Press, 2021.

C.2 Figueres C and Rivett-Carnac T. *The Future We Choose: Surviving the Climate Crisis.* London: Manilla Press, 2020.

C.3 The Guardian. *'Air is cleaner than before the Industrial Revolution': a best case scenario for the climate in 2050.* Available at: www.theguardian.com/environment/2020/feb/15/best-case-scenario-2050-climate-crisis-future-we-choose-christiana-figueres-tom-rivett-carnac [Accessed: December 2023].

C.4 The Guardian. *'The only uncertainty is how long we'll last': a worst case scenario for the climate in 2050.* Available at: www.theguardian.com/environment/2020/feb/15/worst-case-scenario-2050-climate-crisis-future-we-choose-christiana-figueres-tom-rivett-carnac [Accessed: December 2023].

C.5 Stack. Available at: www.stackmagazines.com/ [Accessed: December 2023].

C.6 United Nations. *Sustainable Development Goals.* Available at: https://sdgs.un.org/goals [Accessed: December 2023].

C.7 The Institution of Structural Engineers. *Annual academics' conference 2022.* Available at: www.istructe.org/resources/training/annual-academics-conference-2022/ [Accessed: December 2023].

C.8 M Shed. *Building a Martian House.* Available at: www.bristolmuseums.org.uk/m-shed/whats-on/building-a-martian-house/ [Accessed: December 2023].

APPENDIX D

Bibliography

Throughout this book we have used the metaphor of bookcases to describe systems change; with the shelves of each bookcase acting as different 'levels' within the system. Somewhat fittingly, we have read enough literature in the course of our research for this book, to fill several bookshelves. Some have been referenced in the relevant chapter, while others have influenced our thinking less directly (but just as importantly). The following is a list of the most influential, with a very brief explanation for inclusion:

Do Build
Alan Moore

Alan Moore explores the role of 'beauty' in design, and in his definition of beauty he asks whether something beautiful can also do harm. The ensuing discussion focuses on many aspects of regenerative design.

On Time and Water
Andri Snær Magnason

Provides a different global perspective on the climate emergency and especially the loss of glaciers.

39 Ways to Save the Planet (podcast)
BBC Radio 4

These short podcasts look at a variety of ways we can change current practice. Many of them relate to engineering.

Woodsman
Ben Law

Describes the creation of a house following regenerative practices.

Atlas of the Heart
Brené Brown

In Chapter 4 we discuss the importance of observation. Brené Brown's book explores emotional literacy, which is a key part of observing communities and understanding our own response to what we observe.

The Future We Choose
Christiana Figueres and Tom Rivett-Carnac

This book paints two futures; one where we do all we can to avert the climate emergency, and one where we don't. This is a very powerful way of presenting the possible outcomes that lie ahead.

The Second Body
Daisy Hildyard

Inspired the idea of the 'Second Site'.

Doughnut for Urban Development
Dani Hill-Hansen and Kasper Guldager Jensen

This book takes the philosophy of doughnut economics and applies it to urban development. It contains some helpful guidance on measuring impact across a wide range of factors and considers both local and global impacts.

Designing Regenerative Cultures
Daniel Christian Wahl

This is the definitive text on regenerative design. It is packed full of wisdom and great questions. However, some understanding of systems thinking and regenerative practice will greatly aid understanding of the text.

Thinking in Systems
Donella Meadows

This is a clear and simple guide to systems thinking, and we would highly recommend it as a companion text to this book.

Under a White Sky and *The Sixth Extinction*
Elizabeth Kolbert

Under a White Sky explores three ecological challenges and examines whether we can solve them at the same level as we created them. *The Sixth Extinction* takes a detailed look at the history of biodiversity loss, and explains why this extinction event is profoundly different to the previous five.

What White People Can Do Next
Emma Dabiri

In Chapter 4 (Section 4.1.3) we discuss inequality. This short, helpful guide sets out what we might do to address the issue.

No one is too small to make a difference
Greta Thunberg

This short and inspiring text captures many of Greta Thunberg's speeches on the climate emergency.